Algrove Publishing Limited
1090 Morrison Drive
Ottawa, Ontario
Canada K2H 1C2

Canadian Cataloguing in Publication Data

Bullock, William
    Timber : from the forest to its use in commerce

(Classic reprint series)
First ed. published: London : Pitman, 1915.
Includes index.
ISBN 1-894572-21-1

1. Timber. I. Title. II. Series: Classic reprint series (Ottawa, Ont.)

SD434.B85 2000   338.1'7498   C00-901605-8

Printed in Canada
#11200

PUBLISHER'S NOTE

Anyone who has spent much time in Great Britain will have heard the term "deal" applied to lumber. In this book, it generally applies to lumber of all dimensions from conifers, similar to the North American term "SPF" used to describe a product that could be made of spruce, pine or fir, alone or in combination.

But "deal" had a more specific meaning on both sides of the Atlantic at one time. In Great Britain a deal was understood to be 9″ wide, not more than 3″ thick and at least 6′ long. If shorter than 6′, it was a deal-end. If not more than 7″ wide, it was a batten. If between 2″ and 4″ thick, and between 2″ and 4¹/₂″ wide, it was called a scantling.

In North America the standard deal was 12′ long, 11″ wide and 2¹/₂″ thick. Wood 12′ long, 11″ wide, but 1¹/₄″ thick was called whole deal. Halved again (i.e., 5/8″ thick) it became slit deal.

In view of all this, today's practice of naming lumber by its dimensions seems unquestionably reasonable.

Leonard G. Lee, Publisher
Ottawa
November, 2000

A GIANT PINE, CALIFORNIA

PITMAN'S COMMON COMMODITIES OF COMMERCE

TIMBER

FROM THE FOREST TO ITS USE IN COMMERCE

BY

WILLIAM BULLOCK

LONDON
SIR ISAAC PITMAN & SONS, LTD., 1 AMEN CORNER, E.C
BATH, NEW YORK AND MELBOURNE

PRINTED BY SIR ISAAC PITMAN
& SONS, LTD., LONDON, BATH,
NEW YORK AND MELBOURNE

PREFACE

AT the desire of the publishers I have attempted, in the following pages, to write some account of the timbers which enter so largely into the composition of many objects in our everyday life.

The literature relating to the subject is not extensive, and, with one or two exceptions, the books that have been published have been written more with a view to disseminate botanical or other scientific information than to give a practical knowledge of the woods in every-day use. As this book is for popular consumption, I have, therefore, endeavoured to make it as free from technical and scientific terms as possible, and have also, partly from the limits of the pages, strictly confined my remarks to the timbers which are at the present time dealt with in ordinary commerce, and of which I can speak with a practical knowledge gained in a rather lengthy experience in handling, buying and selling.

For some interesting statistics in reference to a few of the woods, I have to tender my thanks to the Editor of *The Timber Trades' Journal*, and I must also express my obligations for further records which were obtained from the well-known firm of London Wood-brokers, Churchill & Sim.

WILLIAM BULLOCK.

STOKE NEWINGTON,
LONDON, N.

CONTENTS

LIST OF ILLUSTRATIONS

TIMBER

CHAPTER I

INTRODUCTION

FEW products of nature are of more manifold utility than timber and few have such powers of adaptability ; whether our outlook may be in the midst of cities, in the centre of agricultural districts or amid primitive civilization, we see on all sides convincing proof of its general use and necessity. It is with us in some form or another from the cradle all through life to the end of one's existence, and the most casual observer cannot fail to recognise its ever present and near relation to all the numerous commodities of everyday life.

It is therefore natural that a product with which we are all, to a more or less extent, so familiar should form an interesting subject to dilate upon, condensing information which has been gained by long and close connection with the material, in order to afford readers, who have perhaps had their interest centred in other channels, some knowledge of the many varieties of woods that exist, their varied qualities, their uses and the part they have played in the progress of the world.

The greater part of the globe at one time or another was covered with timber forests. The conditions were favourable for reproduction and the result was vigorous regeneration by the creative power of the soil and climate. Gradually man and civilization advanced and reckless attrition of these forests took place, by fires to clear the land for agricultural purposes, by unnecessary extravagance in supplying domestic requirements and other ways. This prodigal waste was rampant wherever civilization progressed, and gradually such tracts of country became denuded. This extinction of the forests which had protected the land led, especially in tropical countries, to radical alterations in the water supplies, to changes in the climate, to great erosion of land and, finally, to a transformation in the sylvan character of the country.

The position at the present time is that many countries are more or less devoid of timber, others by careful afforestation have kept up a moderate supply, while in several more, mostly undeveloped or which are in course of development by civilization, the reckless waste and non-provision for the future, as carried on in bygone times, is still more or less prevalent.

Statistics of the forest areas of many parts of the world at the present time can only be vaguely estimated. Those relating to Europe may however be taken as approximately correct, and Sir William Schlick estimated that in 1901 over 758,000,000 acres of land in Europe were more or less afforested, the figures giving a proportion of 31 per cent. of the total area, and to the population 2 acres per head. The full figures in regard to the different countries in Europe are interesting and from the above author's Manual of Forestry the following details are quoted—

Russia	has	516,000,000	acres afforested		40 per cent. of the total			
Sweden	,,	48,000,000	,,	,,	40	,,	,,	,,
A.-Hungary	,,	46,410,000	,,	,,	30	,,	,,	,,
France	,,	23,530,000	,,	,,	18	,,	,,	,,
Spain	,,	20,960,000	,,	,,	17	,,	,,	,,
Germany	,,	34,490,000	,,	,,	26	,,	,,	,,
Norway	,,	17,000,000	,,	,,	21	,,	,,	,,

Italy	has	10,110,000	acres	afforested	14	per cent.	of the	total
Turkey	,,	6,180,000	,,	,,	8	,,	,,	,,
G. Britain	,,	3,030,000	,,	,,	4	,,	,,	,,
Switzerland	,,	2,100,000	,,	,,	20	,,	,,	,,
Greece	,,	2,030,000	,,	,,	16	,,	,,	,,
Portugal	,,	777,000	,,	,,	3	,,	,,	,,
Belgium	,,	250,000	,,	,,	17	,,	,,	,,
Holland	,,	570,000	,,	,,	7	,,	,,	,,
Denmark	,,	600,000	,,	,,	6	,,	,,	,,
Bulgaria and Herzegovina	,,	10,650,000	,,	,,	45	,,	,,	,,
Bosnia	,,	6,790,000	,,	,,	53	,,	,,	,,
Servia	,,	2,390,000	,,	,,	20	,,	,,	,,
Roumania	,,	5,030,000	,,	,,	17	,,	,,	,,

As to other countries in the Eastern hemisphere, the Siberian portions of Russian territory are known to be well afforested; Japan and Manchuria have also considerable resources, but little is known in regard to China. About a twelfth of the total area of India is under woodland, and estimates of the area on which marketable timber is growing in Australia give a total of over 1,000,000 acres. Little is known in a positive way in regard to the timber resources of the great Continent of Africa, but extensive forests certainly exist in the central portion and along the Northern districts, principally in Algiers, where about 5,000,000 acres of forests are known to be under French control.

In the Western hemisphere Canada has large resources, but with the steady and rapid development of the country enormous inroads are being made on her supply, and unfortunately with no corresponding replacement on a commensurate scale. British Columbia is also a heavily-wooded country; but, fortunately perhaps, has not been exploited to the same extent, the long voyage round Cape Horn, which was necessary to bring the timber to European markets, retarding its shipment; doubtless, on the opening of the Panama Canal, a rapid development in the exploitation of wood from this country will soon be made.

The United States form a typical example of the rapid exhaustion of the woods of a country under the advance of civilization and in conjunction with insufficient care in husbanding supplies and neglect of re-afforestation. The country on its first discovery was practically covered with virgin forests of large extent, but, by fire, reckless waste, commercial greed in marketing all and every sort of timber that could be cut down, and an almost total neglect of replanting, is within sight of a practical exhaustion of its timber supplies. Already it has begun its transition from an exporting country to an importing one, large supplies from Canada and other places being needed to fill its consumers' demand. It is prophesied that twenty years hence no native timber of any kind will be available for export.

Turning from the United States to the various central American Republics we find that most of these States are well timbered and only partly developed. Some, or most, of the West India Islands are depleted or in course of becoming exhausted, no provision so far as is known having been taken to regenerate. Colombia, Venezuela, Ecuador have large resources which have been little exploited. French Guiana is also rich in timber, a similar position being held by British Guiana, although in this latter possession, where there have been signs of failing supplies, conservation has been introduced to a small extent.

The great and comparatively unknown country of Brazil is richly afforested with tropical woods, which are all but unknown, while other extensive countries in Southern America—Peru, Bolivia and the Argentine— are understood to be but sparsely wooded.

It may thus be seen that a more or less rapid exploita- tion of the timber resources of the world is in progress, and that, with extending civilization and increased demands for timber, the consumption is yearly becoming greater. As this demand increases, so the supply becomes less and consequently, year by year, prices, taken all round, show an advance. Speaking generally there is practically little conservation taking place on a commensurate scale, and the prospects are clear as to the future—namely, an ever-growing demand, lessened supplies, and ever-advancing rates.

CHAPTER II

A BRIEF and retrospective sketch of the history of
timber, especially in reference to its uses and the important part it has played in the development of Great
Britain, may perhaps be of interest and form a fitting
introduction to the several chapters which it is intended
to devote to a few of the principal woods which are in
regular use.

It is on record that, at the time of the Conquest in
1066, England must have been of a densely afforested
character, the Domesday Book of 1085 showing that
in the five counties of Derbyshire, Kent, Sussex, Surrey
and Yorkshire alone, no fewer than 1,033 woods
and forests existed. The scattered but rapidly
disappearing remnants of these forests at the present
time are some little testimony to the past wooded
character of the land. To mention only a few in the
Midland and Southern Counties, we have Sherwood
forest in Nottinghamshire, Charnwood in Leicester-
shire, a small and scattered portion of Shakespeare's
great forest of Arden in Warwickshire, Whittlebury and
Salcey in Northamptonshire, Needwood in Staffordshire, Ashdown in Sussex, Waltham or Epping forest
in Essex, and the New Forest in Hampshire. Besides
these may be added the several Chases and Warrens,
such as Cannock, Malvern, Hatfield and Loxley in
Yorkshire, and others which, formerly well-timbered
lands, are now but partially wooded.

Supplies of oak, ash, wych elm, willow, yew and
other trees must have been abundant in these forests,

the first mentioned wood doubtless largely predominating. With the rapidly increasing needs of a growing community these timber resources were freely and rapidly drawn upon up to the sixteenth century. Material for the construction of bridges was needed and for the making of galleys and, later, for the ships by which communication with oversea countries was obtained : much was also consumed for domestic uses, for the building of dwellings, for fuel and many other purposes. With a growing population there also arose a greater need for agricultural acreage, and a consequent clearance and grubbing up of much wooded land took place ; moreover a large supply was needed for smelting, as the increasing use of iron began to arise.

England's naval supremacy, which at this time began to be asserted, not only was due to the inborn fighting qualities of her seamen, but was assisted by the capabilities of her shipbuilders and by the virtues of her native-grown oak. So far-famed were its qualities, that it is most authentically recorded that one of the instructions given to the Duke of Medina Sidonia, who commanded the great Spanish Armada fleet in 1588, was that he should, if he effected a landing in England, destroy the Forest of Dean area, oak from this locality being, it was understood, largely used in the building of English ships.

The increased use of this wood for shipbuilding and other purposes and the first felt difficulty of obtaining supplies was experienced about this time, and a great outcry arose in regard to the prodigal waste that was rampant respecting its use. It was prophesied that its extermination would bring about the ultimate downfall of the English nation, and, as early as 1531, an Act to arrest the depletion was put into force ; another of more important character, which was designed to enforce restrictions respecting the felling of trees and to prevent the conversion of afforested lands into agricultural acreage, being placed upon the Statue Book ten years later. Another enactment in 1558 was also passed, regulating and limiting the size in which wood could be utilized for smelting purposes, and in 1592 a further Act was instituted.

Under the rapid development of the country by the introduction of steam power and its application to the means of transit by land and water, the native supplies of timber became wholly inadequate ; increased demands arose and, to fill these, timber from other European sources began to find its way into the country.

It is recorded by some authorities that, as early as the sixteenth century, fir timber in partially squared condition or balks as they were called, was brought from the Baltic, probably from Dantzig or other Prussian ports, and used on the East coast for masts, spars and other purposes. It was not however until later that supplies increased, when, together with this fir,—red or yellow deal as it is called in our days—oak from the same ports began to relieve the difficulties of consumers in finding sufficient native-grown wood. The imports, both of the above mentioned fir timber and of the oak, grew largely, and then, it is understood, through the enterprise of Dutch merchants, additional supplies of Norwegian and Swedish fir began to arrive. Later, oak and other woods from the acquired Colonial Possessions in America were first brought over.

A new era in the history of the trade thus arrived : no longer was the community dependent on native woods, and, with the rapid development of the country, growing calls were forthcoming for the important

material.   From the outset all this foreign wood shipped
to English ports was subject to a Government duty.
These taxes were largely differential, the imposts having
been greatly in favour of the Colonial products.

They were, however, between 1842 and 1851, gradually brought into unison, and in the latter mentioned
year were made equal by the substitution of an all-
round rate of 7s. 6d. and 10s. per load of 50 cubic ft.
according to description.   In 1860 they were again
reduced to 1s. per load of hewn timber and 2s. 6d. per load
of sawn and, finally, in 1866 they were altogether repealed.

As above noted, all this wood when first imported was
in the nature of hewn balks, round tree trunks roughly
squared by means of the adze, and to convert these into
suitable dimensions for the convenience of consumers pit
sawyers were employed.   These, a sturdy race of men
fitted for such arduous labour, had at this time and up
to fifty or sixty years ago, a halcyon existence, and it
was no doubt, in some measure, due to their continually
growing independence, irregularity and neglect of the
abundant work that was available, that the application of machinery for the conversion and manipulation
of wood was brought about.   Even to this day this
now primitive method of converting timber may be
occasionally met with in remote country districts where
mechanical power is not available.

As is generally known, two men were employed at this
method of sawing, and in connection with them
allusion may be made to the old and popular expression
" top sawyer," meaning one in the first rank or at the
head, this referring to the man standing on the top of
the log, who was in authority and guided the saw to
the chalked line impressed on the top of the log.

The introduction of machinery into the country
for the conversion of timber was followed by its use

in many exporting countries, and gradually the receipt of wood in balks as before described was changed and timber sawn into what is popularly known as deals took its place. These methods of exporting the material grew especially in Sweden, Norway and Russia with the result that, at the present time, consumers receive timber cut into almost every size to suit their requirements, flooring-boards ready to lay down, match-lining ready to put up, and mouldings ready to plant on the work.

The importation of timber, and more especially that of the coniferous species suitable for building construction, grew by leaps and bounds, Swedish, Norwegian, Russian, Canadian and other afforested regions in all parts of the globe were laid under contribution, and its final growth up to the present day may be best brought to the reader's notice by the inclusion of the Board of Trade returns for the year 1914, just concluded.

STATISTICS OF THE IMPORT OF WOOD INTO THE UNITED KINGDOM
DURING 1914, AS ISSUED BY THE BOARD OF TRADE

Sawn and Planed.	*Loads.*	*Value.*
Russia	1,707,030	£5,004,932
Sweden	1,326,753	4,175,630
Norway	262,068	1,022,016
United States	377,084	1,539,560
Canada	847,380	2,812,608
Other Countries	104,820	396,801
Sleepers	215,543	656,000
	4,840,678	15,607,547
Hewn and Pit-props	3,128,648	6,528,148
Staves, Mahogany	463,320	3,201,256
Grand Total	8,432,646	£25,336,951

In the above short résumé of the history of timber,

By permission of the " Timber Trades Journal."

A DUTCH WIND SAWMILL

together with a slight account of the trade as carried on in Great Britain, while the remarks have been general, a more particular application has been directed to the coniferous section, or what is known as soft woods, and it may perhaps be convenient, in this preliminary chapter, to give a brief sketch of the history of two or three of the principal woods which hold an outstanding position in another section of the timber trade, namely, the hardwood branch. These, used in the manufacture of furniture, in boat and shipbuilding, in railway wagon and carriage construction, and in innumerable other industries, form an important part in timber commerce. The expansion in the import and consumption was perhaps not so rapid at first as in that of the coniferous woods, but during the last thirty or forty years, with fresh sources from which supplies are continually arriving, the import has been an increasingly progressive one and consumption has extended as the wealth and prosperity of the country have increased.

In reference to oak, which has perhaps held up to the present time the most important position compared with other timbers which are classed as hardwoods, the early history of the native-grown species, its extensive use in early days, and, notwithstanding the efforts of the State, its gradual depletion have previously been noted. The first sources from which was obtained other wood to augment our failing supply were some districts in the Baltic, Dantzig being probably the first port to send wood. Later, exports were received from Stettin and Memel, and at a near period Russian shipments from Riga and also from Odessa in the Black Sea began to find their way to the markets. Towards 1861 American oak was first shipped to England, that from Maryland and supplies shipped from Baltimore being the first timber of this variety introduced.

For shipbuilding purposes iron had, however, now become established, and in this direction less oak was needed and little growth for a short time took place in the receipt of the above exports from the West.

The finer woods from Memel, Riga and Odessa were however largely used for furniture and building fitments, and that from Prussian ports for railway and rolling-stock purposes. About 1884 the United States began further to supply these markets, the timber they forwarded at this time being practically all in a converted form, usually known as lumber. For the first ten years or so there was a steadily increasing import of this manufactured wood, and later still further expansion took place, the supply during the last fourteen or fifteen years having formed a by no means small proportion of the entire import of this wood. The Slavonian forests of Austria-Hungary have also been laid under contribution, a small share of this fine wood having been continuously shipped to English ports during the last twenty years. A further source of supply has been opened up since the war between Japan and Russia, the first-named country having sent large shipments of logs and converted timber which have found considerable favour.

Mahogany, with which fine and popular wood all are familiar, is understood to have been first introduced into the country about the commencement of the eighteenth century, but it was evidently known among the West Indian Islands as a valuable wood as early as the sixteenth century, Sir Walter Raleigh having commented upon it as a valuable shipbuilding wood. It is generally supposed that the first arrivals came from the Spanish Islands of St. Domingo or Cuba, and

even to this day mahogany of good quality of any
origin is often described by the uninitiated as Spanish
mahogany, although shipments from the first-mentioned
Island practically ceased many years ago. The wood on
its introduction was doubtless in great favour and supplies from St. Domingo in considerable amounts
regularly followed. Other districts in the West Indies
and Central America in which varieties of the wood
grew took up the export, and very soon this tropical
product was being shipped from Mexico, Honduras and
Cuba. Later, other countries in the same zone assisted
in augmenting the supply until, at the present time,
almost every district in Central America and adjacent
countries helps, with wood of varying description, to
fill the ever increasing demand which exists in very
many countries.

A marked epoch in the history of this wood, as these
Central American supplies began to show slight signs
of exhaustion, was the discovery that, in the tropical
regions on the West Coast of Africa, lying somewhat
in the same parallels of latitude, huge forests more
or less timbered with species of mahogany and other
trees were available for supplementing the world's
demands. These enormous belts of forest land, lying
roughly speaking at a distance of about forty miles from
the coast, and extending in a more or less parallel direction to the sea-board, stretch approximately from the
French Guiana possessions to within the country beyond
the Congo, embracing in this extent Liberia, the French
possessions on the Gold Coast, Nigeria, the Cameroons,
the French Congo adjoining, and the Portuguese possessions South of the Congo. Various species of timber
from these forests found their way to the markets in
a desultory manner during the latter part of the nineteenth century, but no accurate statistics are available

as to the quantity or description : there was probably
however little if any mahogany among them, and it was
not until about 1891 that any serious exploitation of
the mahogany supplies of these districts was made.
At first this mahogany was received in Great Britain
with reserve, and for some time considerable prejudice
hampered its sale when in competition with established Central American varieties.   It rapidly grew in
favour however and, assisted by a strong demand which
sprang up for the wood in the United States, the import
and volume of business which followed grew considerably.   How the export expanded, as compared with the
amount of wood brought in from the older sources in
Central America, can best be brought before the reader
by the subjoined tables, which give not only statistics in regard to the United Kingdom, but figures
relating to its import to other European countries.

Teak, another important hardwood that has held,
and still holds, an almost unassailable position for
various purposes may also be referred to.   This invaluable timber, whose native habitat is East India, Siam
and adjacent countries, came into general use, so far
as records show, about the middle of the last century.
Obtainable in long lengths and of fine dimensions and
being found moreover to possess the virtue of rendering
iron, when in contact, immune from rust, Teak rapidly
found favour among ship and boat builders and was
often found to be valuable for railway work.   It stands
in a class by itself in Lloyd's list of shipbuilding woods
and has well sustained its reputation throughout its
career, no wood having yet been found to supplant it
in its use for special purposes.   Owing to difficulties
of extraction and the care with which it is conserved
by the Indian Forest Department in Burma and other
parts of India, the supply has rarely exceeded the

Import of Mahogany Logs into Europe During the Year 1913

Port	Hon-duras	Tobago	Minatit-lan, etc.	Nicaragua Guatemala	Panama	Costa Rica	Colombia	African Sundry Ports	Gaboon	Cuba	St. Domingo	Total
	Tons	Tons	Tons	Tons	Tons	Tons	Tons	Tons	Tons	Tons	Tons	Tons
London . . .	11,407	595		2,537	10	507		12,694	9,184	5,216	19	42,169
Liverpool . .	257	1,535		1,000	387			79,088	3,974	5,876	133	92,250
Glasgow . .	239	100						56	5,424	73		5,792
Other Ports in U.K.										173		273
Total to U.K. .	11,903	2,230		3,537	397	507		91,838	18,582	11,338	152	140,484
Germany . .	3,300	1,730				135		22,250	108,480	520	150	136,565
Holland . .	102	304						11	24,506	1,180		26,103
Belgium . .	42	187								2,670		2,899
France . .	379	4,230	1,107	66	48			14,406	10,781	7,065	220	38,302
Other Countries		100										100
1913	15,726	8,781	1,107	3,603	445	642		128,505	162,349	22,773	522	344,453
1912 . .	14,506	14,452	1,145	445	821	60	49	84,565	105,394	24,695	879	247,011
1911 . .	14,970	12,550	188	1,341	137	35	885	68,577	97,921	24,230	1,238	222,072
1910 . .	12,711	15,276	25	1,384	21	68	580	66,270	66,681	16,649	523	180,188

demand and it has, especially during later years, greatly appreciated in price, the value of the wood at the present time being far in advance of that of mahogany. Most of the supplies are obtained from Burma, but equally fine wood is shipped from Siam and small supplies of an inferior description are also exported from Java. More particular notes of the various descriptions and qualities of these varieties are given in a later chapter which deals with the supply of this and other woods from India and other Asiatic countries. Subjoined, however, are some statistics which give some idea of the import which has taken place during the last five years.

STATISTICS OF THE IMPORT OF TEAK FROM VARIOUS PORTS INTO LONDON DURING THE LAST FIVE YEARS

Year			Moulmein loads	Rangoon loads	Bankok loads	Java loads	Total loads
1914	.	.	76	3,585	1,792	506	5,959
1903	.	.	339	5,671	1,111	593	7,714
1902	.	.	56	3,460	987	894	5,397
1901	.	.	268	4,203	3,217	2,012	9,700
1900	.	.	397	3,190	1,486	1,832	6,905

In the foregoing brief outline of the history of timber a more particular narration of facts and events in regard to its connection with England has perhaps been observed. The history might have been given greater scope and enlarged more fully upon the forests and timber resources of other parts of the world, their size, methods of exportation and other interesting matter, but the limits of this small book preclude such an exhaustive survey, and the history has consequently been limited to the salient facts which relate more especially to the British Isles.

Following the preceding introductory review, the remaining part of the book has been separated into two divisions, one dealing with the coniferous or needle-leaf variety of trees, which are commonly known by the timber trade as " soft woods," the other giving some account of the broad-leaf description, most of which, in a general sort of way, are classed by the same authorities as " hardwoods."

In attempting to give some particulars of the woods in these two divisions, it would be impossible and at the same time unneedful to include every known or partly known wood. Large numbers are only recognised by native or falsely misleading names, many are unidentified and their value, if any, partially or altogether unknown. No mention of these has been made and attention was confined to a moderate list, which it is believed, however, comprises all, or practically all, the timbers which can be met with at the present day in use among consumers.

CHAPTER III

UNDER the name of Conifers or Cone-bearing trees
are included the firs, pines, larches and cedars. The
number of varieties of the various species is arge, and
the forests in which they grow, while considerably
more extensive in the Northern hemisphere than in
the Southern, are well distributed throughout the
entire world, their area being probably greater than
that covered by the broad-leaf section. They occupied
a most prominent position in the area of forests in
prehistoric times, as from fossil remains we learn
that, besides the many species known to us at the
present time, many additional members of the same
family existed.

This wood is more generally used than any other,
being the material most adapted for building purposes
and use in almost every other branch of industry where
wood is required. Records of its use in bygone times
are plentiful : one instance as an example may be mentioned, the mummy cases in the Egyptian room, which
attracts so many visitors at the British Museum,
being constructed of one species of this family,—the
cedar of Lebanon.

Great Britain is by far the largest importer of these
coniferous woods ; and to gain some idea of the magnitude of this trade one has only to turn to the statistics of the Board of Trade which are here included. It
is shown in these that Russia, Norway, Sweden, and
other Northern Countries are the chief sources from

which the supplies are drawn, but others in the Western hemisphere also contribute to the total—Canada, the Southern States of America, British Columbia and other countries.

Considerable confusion has grown up and become established in very many countries in the nomenclature of this species of wood. They are all of the pine family, but names have been loosely applied, and for those not intimately associated with the buying, selling and use, much confusion is engendered. The popular conception in England is, however, to call most of them by the general name of deal, although even there alternative names for the same wood are common, the wood, of the same description as the Scotch pine or fir, so largely imported from Northern Europe, being defined as yellow deal in the Southern parts of the Kingdom and red deal on the East Coast, in the Midlands and further North. As this is decidedly the most important of the coniferous woods, the opening notes on the various varieties may well be commenced with this timber.

Northern Pine.—Shipped as redwood from Russian, Swedish and Norwegian ports, while similar timber from Prussia is defined as Baltic redwood. It is the ordinary red or yellow deal of commerce, so much in use in the building trades and common everywhere in Europe. An allusion was made to this wood in the brief introductory retrospect, its history in Great Britain being traced from its first import from Prussian ports as spars and masts, the growth of the import from these Baltic ports and the development of fresh sources of supplies from Norway, Sweden and Russia.

Principally with this coniferous variety of wood the countries of Northern Europe are, as has

By permission of the

"Timber Trades' Journal."

DEAL YARDS, RIGA

been noted, largely and densely afforested; and, although the consumption during the nineteenth century was exceedingly heavy, Norway and Sweden are the only two which show signs of exhaustion. Russia has however supplied increasing quantities, and the boundless expanse of this country, densely wooded and in many cases with virgin forests that still await the woodman's axe, warrants the belief that no shortage of this description of timber will occur for many years to come. There are, however, authorities who warn us that a limit of time for the producing capacities of this great country can be approximately fixed, and that the supply, while at present abundant, is by no means so large as is generally supposed. A drain on these timber resources, additional to those made by the United Kingdom, comes from France, Holland, Belgium, Germany and other European countries, and also from many British Colonies—Australia, Cape Colony and others. The botanical name of the tree from which this commonly known wood is obtained is *Pinus Sylvestris;* and it is, as stated, identical with the Scotch fir. Grown however in dense forests it is straighter and cleaner in growth than those usually found in Scotland, the trees, grown in such close proximity, attaining a height of between 100 and 150 ft., frequently without a single branch from the base up to about 80 ft. or more.

The amount exported in the log or balks, as it first came to Great Britain, is at the present time exceedingly small—trivial—in comparison with that shipped in a converted state. Planks, deals, battens and boards, ranging from 4 in. thick and from 11 in. wide down to small sizes such as $\frac{3}{4}$ in. $\times 1\frac{1}{2}$ in. or similar dimensions suitable for slating and other purposes, are exported in enormous quantities from the various ports, besides much fully-dressed material, principally from Sweden, consisting of floor-boards, and match-lining ready for the carpenter's use, and mouldings of every description for the use of the joiner.

The timber is of yellow or yellowish-red colour; that with a more pronounced reddish tinge being grown, it is generally understood, on high land, and being of more resinous nature is more favoured for outdoor work where durability is an object.

In exposed, outside situations, the timber is durable when free from sap, and lasts fairly well when in contact with the earth. It is one of the strongest of the coniferous family and the principal timber used for roof timbers, joists, window-frames, doors and general joinery. Most of the best timber is shipped from Archangel and other ports in the White Sea, but deals and timber of almost equally good description are also supplied from Petrograd and some of the Swedish and Norwegian ports.

The wood, after chemical treatment, is also largely used in Great Britain for railway-sleepers, and large shipments for this purpose come, principally, from Baltic and Russian ports.

A large export of short lengths of the timber for pit props is also made, these goods being mostly directed to the North-East Coast of England to supply the wants of collieries in the Durham and Midland areas. A great export of deal and other short ends takes place, this wood being principally used for the making of small packing-cases and for firewood. Sweden also ships much timber in the form of poles, which are used for telegraph purposes and for builders' scaffolding.

White Pine.—This, another European member of the coniferous family, is commonly known as white deal or Baltic whitewood. It is one of the hardiest conifers, and exists at a height of about 3,000 ft. above sea-level ; the best specimens, however, attain their greatest size in the valleys. It is found in the same regions and shipped from the same ports in much about the same manner as the Northern Pine, excepting that the qualities exported are extremely limited in comparison with those of the latter species.

It is one of the loftiest of European trees, often growing to a height of between 80 and 100 ft. ; it is however not so quick growing as the Scotch fir. Unlike this last-named variety, it has a development of its branches low down towards the base of the stem, assuming in its growth a somewhat pyramidal form.

The wood is of white or yellowish-white colour, tough and elastic in fibre, but not so strong as red deal. It works fairly well, having in good qualities a satiny lustre after the process of planing, and, although much of the timber has the defect of having number-less small live knots in its growth, it forms an excellent material for inside joinery, for which it is principally used. It holds however a greatly inferior position and is much less valuable than red deal owing to its want of durability when exposed to the weather. The best qualities are found in the shipments from Petro-grad and Archangel, those from Riga and other Baltic ports being of somewhat coarse fibre. Besides its use in the building trade for inside purposes, it is in favour with the cabinet maker for kitchen table-tops, and to some extent for other purposes, owing to the great appreciation of late years in the price of Quebec pine. The poorer qualities are largely used in the packing-case trade, and there is a large export to Great Britain of floor-boards and match-lining which are principally used for cheap house-building. Great quantities are also used in the manufacture of paper pulp, the export

By permission of the "*Timber Trades Journal.*"

FIR TIMBER FROM NORTHERN FORESTS. READY FOR THE MILLS AT RIGA

of which from the various countries, but principally from Sweden, shows an ever increasing volume. The thinnings of the woods are also exported and used for scaffold poles, pit-props and other purposes.

Larch.—This is another species of the coniferous family, widely distributed in Northern Europe and also freely grown in Great Britain. A moderate amount of timber of this species is imported thither, principally from Russian ports, from whence the finest qualities are shipped, further supplies arriving from the Baltic.

Being grown in most districts of Great Britain the timber is well known. It is of a very resinous nature and one of the most durable of the fir or pine family, the better qualities even holding their own in competition with oak, where used in contact with the earth. The wood is of reddish-brown colour, clean and free from knots, but more difficult to work than red deal. Owing to its tough and durable qualities it is used for floors, stairs and other purposes where wear and tear is heavy, for boat-building, posts and rails, rustic work, small scaffold poles, and largely for pit-prop purposes.

Many other varieties of coniferous timbers that are grown in European countries might be mentioned, but they hold a relatively minor position in comparison with those given. One variety, however, may be alluded to, as a large amount of the wood of this species is imported into Great Britain in the shape of pit-props. This is the Cluster Pine, of which large and scientifically cultivated forests exist in the Landes and Gironde districts of France. The wood is used to some extent in carpentry work but is not of very much account, the chief value of the trees being the use which is made of them by the extraction of resin, from which the turpentine of commerce is obtained. A fairly large industry is carried on in the above districts, and the trees after exhaustion and also the thinnings of the forests are largely exported, principally from Bordeaux for pit-props. The chief supplies for the coalfields of South Wales are drawn from this source.

From Eastern countries, many of which are practically untapped as regards coniferous timbers, there is but a small import of soft wood. Doubtless there are ample supplies, but while there are the European and Asiatic forests of Russia and the present resources of Canada, British Columbia and other countries comparatively near at hand to draw upon, the question of obtaining material from such distant regions as those in the East is hardly considered.

One failure of supply which is apparently in sight;— namely, the diversion of much of the pine wood of Canada to the nearer United States,—has been seized upon by the shippers as an opportune occasion to export the supplies of the forests of Eastern Siberia. Shipments of excellent pine, of capital specification and well converted, have reached the markets from the distant port of Vladivostock, and the wood, although perhaps not so mellow as the Canadian, is steadily growing in favour with consumers.

Although this is not in the Eastern hemisphere, a few New Zealand coniferous woods may, perhaps, be included in this chapter, since some of them are in fair use in Great Britain.

Kauri Pine. —Although meeting with a poor reception when introduced into England from New Zealand, about 1888, this wood has been steadily growing in the estimation of users ever since, and there is, at the present time, a regular and constant consumption.

The tree from which the wood is obtained is, perhaps, the most important in the Islands, as, besides the useful timber, a gum which is exuded from the tree, and which is largely exported for the manufacture of varnishes, is produced and forms an important article of commerce.

The colony was doubtless heavily timbered with these great trees in its early days but, unfortunately, forest fires, many probably started with a view to clearing the land for settlement, and ruthless waste in other directions, have so depleted the country that, with a great demand springing up for this wood from Great Britain and the neighbouring sister colony of Australia, a visible exhaustion of the supplies is in sight.

The tree grows to admirable proportions, many butts in the shipment that first arrived—a sailing cargo of round logs—having been up to about 7 ft. in diameter. These, as noted, were unsquared logs, but all consignments since received have been in the form of converted planks or boards, exceedingly well sawn and of good specifications.

The wood is of whitish-brown colour, fine, smooth and silky in texture, and is notable for its even-wearing properties and—for a pine—its general durability. It is practically free from defects and highly esteemed as a material for many purposes, but particularly for the deck planks and fittings of yachts and other boats.

New Zealand Pine.—This, another wood of New Zealand growth, was introduced to the notice of consumers in the United Kingdom somewhere about 1900—1912. The tree is known as Kahikatea in the colony, grows to good proportions, and supplies a useful wood.

It is yellowish-white in colour, of soft but firm and even texture, splits very readily, and if not properly cut and seasoned is liable to warp, twist and become discoloured ; it is moreover not at all durable for outside purposes.

TRANSPORT OF TIMBER IN AUSTRALIA

Shipped in well-converted planks and boards, similar to the manner of Kauri, it is, when well seasoned, free from all defects and has found a fair number of consumers. It is largely used in the colonies for packing-case making, and the wood may be seen in use if the boxes, in which New Zealand butter reaches the British market, are examined.

Rimu.—Known in the colonies of New Zealand as " Red Pine." This is another wood which holds an important position in the Islands. Very little has however, been exported, and it is not sufficiently known to have become an established article of Commerce. The timber can be obtained in long lengths and up to 24 in. or more in width, is deep red in colour, and occasionally has a certain amount of figure. It is of fine straight grain and would probably work in a satisfactory manner. For building purposes, joinery and cabinet work, it is in general use throughout New Zealand.

Totara.—Next to Kauri this coniferous wood shares with Rimu the most prominent position as a general utility wood among the native timbers of the Islands.

Similar to Rimu, it is of deep red colour, clean and straight in the grain, durable both for outside and inside purposes. It is procurable in long lengths and wide widths, and is used in the colony for general building purposes, joinery and cabinet-making, bridge construction, paving and other work. There has been practically no export to the United Kingdom.

CHAPTER IV

THE CONIFEROUS TIMBERS OF CANADA AND OTHER COUNTRIES IN THE WESTERN HEMISPHERE

Weymouth Pine.—A well-known and popular timber which is known as yellow or Quebec pine, has been exported from Canada since about the year 1705. It is one of the most important trees grown in Canada, and large areas of that country are covered with its growth. The nearer forests have; however, been greatly depleted by the large export trade that has been so long in existence, and the cost of exploiting the forests further afield has considerably enhanced the value of the wood during the last ten years. It was also formerly plentiful within certain latitudes of the northern part of the United States, but that supply has rapidly reached an end by the demand of local and near consumers. Ornamental specimens of this tree may be seen in many districts in the British Isles, and growing to fine proportions both in height and girth of trunk it is one of the most attractive conifers that is grown.

The timber, when first brought into Great Britain at the beginning of the eighteenth century, was introduced in the shape of hewn logs and found much appreciation. It was used extensively for deck planking and interior fitments mostly among shipbuilders. The import of logs in this form has continued up to the present time, but to nothing like the extent as formerly, and, known as waney board pine, they are still used to some extent for similar purposes. The bulk of the trade however, after the introduction of machinery, developed in the same way as mentioned in the remarks concerning the changes in the imports from Northern Europe.

Converted timber, mostly cut to 3″ by 11″ and over were exported, and a very extensive trade, principally with Great Britain, grew and expanded.

Up to about 1889 most of these sawn goods were floated down the rivers of Canada to the port of shipment in huge rafts, and it was customary to grade those planks or deals that had been immersed at the bottom of the pile as " floated goods," those at the top as " bright," a difference of 10s. per standard showing the estimate of their values. This system lapsed as other and better means became available to convert the logs nearer port of shipment. At an early period all these converted goods were graded into firsts, seconds, and thirds ; later, when near forest supplies began to show signs of failure, fourth quality were added ; and later still an intentional lowering of all grades by shippers was made. A period arrived, about 1890, when the exhaustion of this timber in the northern part of America obliged the consumers in that country to have recourse to Canadian goods, and gradually these nearer markets attracted the Dominion wood. Imports to Britain then began to get scarce, a great appreciation in prices took place and consumption was greatly curtailed. Values have been on the up-grade for the last ten or fifteen years, being at the present time something like 50 per cent. in advance of prices current about the year 1900.

The wood is of yellowish-white colour, turning somewhat brown with age and exposure under natural conditions. It is of soft but firm and even grain, fairly free from knots, works exceedingly well and will not, when free from heat, buckle or twist. It is not at all durable for outside work, neither is it a strong wood, but for interior joinery fitments, ground work for veneering on, coach panels or similar work, it is a favourite wood. It was formerly, when more moderate in price, a staple wood for the pianoforte trade, and its non-casting character, mellowness and general reliability have always made it the ideal wood for the making of patterns.

Red Pine.—This is known as Norway Pine in Canada, and Yellow Pine in Nova Scotia, and is a conifer growing up to about 80 ft. in height and about 2 ft. in diameter. The wood is of reddish colour, generally clean in growth, and possesses good working qualities. It is durable and will stand well in outside positions, being, in most respects, similar to European red deal. Imported into England about the year 1756, in the shape of logs suitable for mast and spar purposes, it was much in favour for many years, but owing to the competition of European woods gradually lapsed in use. It is now, and has been for a long time, shipped in the form of planks and deals. Most of these imports arrive in the Thames, the wood having a limited sale otherwise than in the Metropolis and the southern parts of the Kingdom.

Spruce.—This is another cone-bearing tree from which is obtained timber bearing some resemblance to the white wood imported from Northern Europe, and of which some account has been given. It is abundant in the Canadian forests, and is exported in considerable quantities to very many markets.

In the Dominion and also in the States the timber is largely used for building purposes, but in England its use is principally confined to the packing-case trade, although in the Midlands and North-East a good proportion of the imports are used by builders for joists, rafters and other constructive parts of house-building.

The wood is an excellent material for floor-boards, and much used in the Midlands for that purpose where a strong rough floor is a necessity, as for instance in factories. A fair amount is used by the pianoforte trade, " bracings " being always constructed of this wood. It is generally imported in 3′ by 9″, 3′ by 11″ and smaller sizes, that shipped to the London market being generally in 12 ft. and 13 ft. lengths. The wood is white in colour, tough and strong, does not work very kindly, but planes well and leaves a clear satiny surface. It is durable both for outside and inside work, but is rather inclined to twist, and in the inferior qualities has, occasionally, many dead knots. The deals or planks exported are usually graded, those arriving for the London market being generally divided into four classes.

Columbian Pine.—Alternative names for this timber, as with many others, are common. It is known as Oregon Pine, Yellow Pine of Priget Sound, Douglas Fir, yellow or red fir in the United States, red fir or red pine in Canada. It is the chief tree of British Columbia, and from the heavily afforested valleys and hillsides of this far-away territory the largest exports are made from Vancouver and other Pacific ports. It is, however, also found in Canada and growing on the slopes of the Rocky Mountains in California, reaching its most perfect development in the coast regions of Washington, Oregon and British Columbia.

The tree is said to attain a height of between 200 ft. and 300 ft., with a diameter up to 8 ft. The wood is of strong nature, durable, somewhat coarse in texture, and difficult to work. With a tree of such fine growth, timber of excellent specifications can be converted, and the wood as a constructive material has a great future in view. It is held in high esteem for bridge building and similar work, for which it offers the combination of lightness, stiffness and strength, besides furnishing pieces of exceptional length. Squared planks or logs 24 in. in diameter and up to 100 ft. long

Photo by   *Darius Kinsey, Seattle, Wash.*

SHINGLE BOLTS ON A CALIFORNIAN RIVER

are procurable, and it is said can be obtained up to 200 ft. in length. The consumption has shown considerable increase in England, even with the heavy freight charges incidental to its having to be shipped right round the South of America, but with a lessened cost in prospect, when the Panama Canal is completed, it may be predicted that increased sources of consumption will speedily follow.

Sequoia.—Only two living species of this coniferous tree are now supposed to be existent; fossil remains, however, disclose that over forty additional species have at earlier periods been common in various parts of the world. Both the surviving members of the family are found in California, and particulars of their growth are so full of interest that a rather extended account may perhaps be given.

Dealing with the one commonly known as Redwood, which is the variety in commercial use, it is understood that its growth is restricted to a belt of coast-land bordering the Pacific, some 450 miles in length and about twenty miles in breadth. The trees reach their most perfect development in Humboldt County, where virgin forests, which may be traversed for over twenty miles, are monopolized by these great trees.

Said to be very fast growing for the first thirty or forty years, they add perhaps 2 ft. per year to their height during this period, while in their early age, between four and ten years, they possibly grow 2 ft. 6 in. per year. They attain, at mature age, a height of 100 to 340 ft.

The tree sometimes reproduces itself from seed, but more usually, and when growing in thick forests, by the growth of stump sprouts, this being the only coniferous tree which has this habit. Huge lateral roots spread

near the surface, and from these a second generation
rises up.   In some instances these have been cut and
then another encircling generation has come into
existence.   One of these circles is about 50 ft. in
diameter and contains forty-five trees, the growth of
which ranges from 2 ft. to 6 ft. in girth : these are the
third succession, and in the centre stands the parent
trunk.   This shaded enclosure affords sufficient sitting
accommodation for a congregation of about 250 persons,
and here Divine Service has been held.

Some idea of the amount of timber in these giants
of the forest may be gained by the perusal of a book
issued by the Red-wood Manufacturers Association.
In this they mention that in some forest areas in the
Humboldt districts 2,500,000 ft. board measure to the
acre may be found, while from one tree only 480,000 ft.,
not including waste, has been obtained.

The wood extracted from these trees is soft in texture, very straight in the grain, and has easy working
qualities.   It is undoubtedly most durable, posts
buried in the earth having been known to last thirty
or more years, but its great and serious defect is its
brittle nature and want of strength.

It has always been shipped to Great Britain in the
shape of well-manufactured planks of extremely good
dimensions, has been well tried for many purposes
but has not obtained any very large patronage owing
to the defect of its lack of strength.

Big Tree.—This further species is rarely exported,
and no shipment, so far as is known, has reached the
English markets.

It grows to a height of 150 ft. to 325 ft., and from
80 ft. to 225 ft. is reached before branches occur : in
girth it attains a diameter of from 5 ft. to 30 ft. measured
6 ft. from the ground.

Restricted to definite localities called groves it grows at an extreme altitude of 7,500 ft. in regions where the average rainfall is 45 in. to 60 in. per annum, and where the snow lies from 2 ft. to 10 ft. deep during three to six months in the year.

As has been remarked, timber of this species has never been exported, the wood, besides being even more brittle than red-wood, is less readily worked.

A few notes relating to some of these Big Tree Groves, which are a great attraction to visitors to California and which, probably, rank high among the many wonderful sights to be seen in the United States, may perhaps be interesting.

In the Calaveras County, the grove bearing this name contains 101 trees, besides several fallen ones. Thirty trees in this collection, from actual measurement, range from 9 ft. to 19 ft. 6 in. in diameter at 6 ft. from the ground, and rise to a height of 237 ft. to 325 ft. One of the fallen trunks has been hollowed out by fire, and so huge is the tree that a mounted horseman can ride through this aperture.

The Tuolumne Grove is situated at an altitude of 5,800 ft. There are thirty trees in this stand, and an opening has been cut through one, under which the Yosemite stage-coach passes.

A third group—Mariposa Grove—contains the largest tree that is known ; it measures 93 ft. 7 in. in circumference at the ground and 64 ft. 3 in. at 11 ft. above.

The estimate of the ages of some of these giants of the forest shows that many seasons have passed over their heads. A count of the rays has demonstrated that the adult trees have a growth of between 400 and 1,500 years, but one patriarch has been discovered whose annual rings show that 4,000 birthdays must be credited to him.

The excellent cross-section of one of these trees which, exhibited at the Natural History Museum at South Kensington, is so well known to frequenters, impresses one vividly with the idea of the age of these mighty trees. It shows that the tree from which it was obtained was cut down in 1892, its age at that time being 1,335 years. Notes on historical matters are made on this cross-section at the various periods of the tree's growth, and it may be observed that it came into existence about the commencement of the Middle Ages, was in the prime of life at the time of the accession of Alfred the Great, and continued through all the events and progress of civilization almost up to the end of the reign of Queen Victoria, when the woodman's axe brought the giant to an untimely end.

Sugar Pine.—An important Californian timber which is put to many uses, is obtained from a tree bearing this name. It is chiefly employed in the Western States however, comparatively little finding its way thence into other countries. It is most abundant on the western slope of the Sierra Nevada, growing at an altitude of from 2,000 ft. to 7,000 ft. The trees attain to a height of between 80 ft. and 250 ft., with a girth of 2 ft. to 8 ft., and it is said that no other pine is so valuable for its wood or will show such a yield per acre. The demands on the forests for home consumption have been great, and the wood has steadily appreciated in value. It is light, soft, close-grained, and fairly strong, works easily and is fragrant in smell. Like Weymouth and other pine the wood, when converted, sweats and stains if packed together when green. It is said to have been discovered by D. Douglas a noted botanical explorer, who named the species *Pinus Lambertiana* after his friend Lambert, a founder of the Linnean Society.

Port Orford Cedar.—This is another important tree of Western American origin ; its acreage however is somewhat limited. It attains to a height of from 125 ft. to 150 ft., and a diameter near the ground of 4 ft. to 12 ft. The wood is straight and close grained, soft, elastic, durable, and of a yellowish-white colour. It is used for flooring and similar work, and is held in much esteem by Californian consumers. Very little has been exported as yet to England. It was introduced into Europe in 1854, and is known, botanically, as *Cupressus Lawsoniana*, there being now, it is stated, not less than sixty-eight garden species under cultivation.

Pencil or Red Cedar.—This is a species of Juniper tree, widely distributed in the United States, but growing more freely in the southern parts, especially in Florida. It is also met with in Canada but only in limited quantities. The tree is of moderate growth, and yields the wood which has such an extensive use among pencil manufacturers. For this purpose it is unequalled, being of fine, straight, silky grain, easily worked and fairly strong, and no efficient substitute has as yet been found in any part of the world to supplant it in its use for pencil-making. Enormous inroads have been made on the supplies to meet the ever-growing demands for the above purpose, and the exhaustion of the wood is understood to be well within sight. It arrives in small, short logs, which, indicating the difficulty of procuring supplies, show yearly a deterioration in quality.

Pitch Pine.—A well-known timber that has been largely used in Great Britain during the last eighty or 100 years has next to be noted.

Known in the United States as Long Leaf Pine, the above-named timber is, perhaps, the most important in the Southern States of America. Formerly, unsurpassed forests existed, a belt of country stretching some 150 miles parallel with the coast of the Southern parts of North America. Ruthless destruction has, however, laid waste much of this primitive forest, the damage not being due to the need for timber supplies, but to the exploitation of the turpentine industry. A large part of the turpentine of commerce is obtained from the resin extracted from the trees of these forests. British imports alone of this commodity amount in value to something like one million pounds sterling per annum. The trees are highly resinous, but under the severe tapping they are subjected to are quickly exhausted and fall to the woodman's axe; and at such a pace are clearances made that it is predicted that a total extinction of these vast forests is within measurable distance. The timber is principally shipped from ports in the Gulf of Mexico, Galveston, Mobile, Pensacola or New Orleans, and arrives in the shape of hewn and sawn-sided logs—mostly the latter, besides planks and boards.

It is largely used for any constructional work requiring strength, and for any purpose where a timber adaptable for use under entirely wet conditions, or alternating wet and dry, is required. It is not, however, a good material for entirely dry situations, becoming brittle in time. Having a handsome appearance when varnished it was commonly in use in former years for furniture, the fitments of offices and other joiner's work, but it was not an ideal wood for the purpose, being subject to shrinkage and expansion with alternations of the atmosphere. Formerly squared logs of 14 in. or more, and ranging up to 80 ft. or more in length, were common, but the specifications have largely declined of late years, and logs of even 60 ft. in length are not common.

The wood is in moderate use for flooring purposes in England, and in the States is used, with other timber of the same species, for street-paving purposes. The principal use, however, in European countries at the present time is for piles, beams, bridges, and other heavy constructional work.

Loblolly Pine.—This is a timber in many respects identical with the foregoing. It grows freely in the Southern States, forests being plentiful in Virginia, South Carolina and Texas. It is obtainable in better dimensions, but not in such long lengths as Pitch Pine, and is inferior in many respects, having much more sap-wood and being of a soft and coarse texture. Similar in general appearance and of less value, it is often substituted for Pitch Pine and also for the Short-Leaf Pine of Missouri and Arkansas, which is also known under the name of Carolina Pine. No shipments of the wood are made to Great Britain, but it is largely used in the Southern States for sleepers, bridges and piles, after chemical treatment.

Carolina or Short-Leaf Pine.—Another timber of similar resinous nature, common in the Southern States of America, where it grows to a height of 80 ft. It is not the equal of Pitch Pine in strength, and is not so resinous, but in other respects it will bear comparison. Shipments were made to England some years ago in an attempt to establish a market, and it sold fairly well for a time, a moderate consumption taking place in the timber or converted boards which were sent. The demand however subsided, and it is now but seldom seen in the English markets.

Cypress.—A further timber from the Southern States of America, for which attempts have been made to secure a regular and established demand in Great Britain. The tree is allied to the red woods of

California and, together with those forests, occupied in former ages most of the hills and valleys which extended far north. Submerged forests of these trees have been found in the swamps of rivers running into the Gulf of Mexico, and specimen logs, with an estimated age of 10,000 years, have been discovered in the New Orleans Drainage Canal, 18 ft. below the present sea-level of the Gulf. It is essentially a swamp tree at the present time and grows, more or less, in water. In many respects it is similar to the red-wood of the West Coast, having however considerably more strength than that Californian wood.

Its chief qualities are its clean growth, durability, and the absence of any chemical in the wood that will impart odour or colour. Having these latter qualities, it is a good material for tanks, vats, etc., for which it is largely used in the States. The timber is shipped to market in well-converted planks and boards, but has not made very rapid progress as yet in the estimation of British consumers.

An abundant quantity of other coniferous species, of more or less value, which grow in Canada, the United States, Mexico, and other countries in the Western hemisphere, might be mentioned, but the limits of these pages preclude such a list, and attention has, therefore, been confined to those only which are imported and commercially dealt with in the English markets.

CHAPTER V

THE HARDWOOD OR BROAD-LEAF SECTION OF TREES

AMONG this extensive and widely-dispersed species of timber, the varieties of oak and mahogany are so numerous and important, that separate chapters are devoted to particulars regarding them, the observations on other woods being grouped under the heads of the several countries from which they are obtained.

In this present chapter the first-named timber, as one that is always regarded as typical of England and its breed of men, and which has played a not unimportant part in the foundation of her supremacy of the seas, will be considered.

Referring at the outset to English-grown wood, three species are said to exist. The divisions however are more botanical in their definitions than in any divergence in the quality of the timber. The wood certainly varies, but such differences depend more on soil, locality and other influences, than on any inherent distinctions in the variety.

The oak is indigenous to Great Britain and Northern latitudes, extending in fact as far north as 35 degrees. It is also found as far south as tropical Africa. In England it is so well known that it needs no long description. It grows preferably in rich, well-drained loam, and, probably, Sussex, Gloucester, and some parts of the Midland Counties produce the finest timber. As a hedgerow tree it develops many branches, and the timber, wanting in the length of the butt, grows tough, knotty, and is therefore difficult to work. It is, however, useful for many purposes, especially where sweeps and bends are required, as for boat-knees and other purposes. The finest timber is developed when the trees are coppice-grown—straight, long, and clear trunks being the most advantageous form for general use. Unrivalled as are its qualities for durability and strength, and indispensable as it was in times past, it has not, during the last century, been so necessary to consumers owing to importations from other countries. These have partly taken its place, some for their better working qualities, others for their superior dimensions, but for durability and strength the British grown wood has not been equalled, either when wholly submerged in water or subjected to the alternations of wet and dry. Its use of late years has been principally confined to boat or barge-building, wheelwrights' use and, in a much less degree than formerly, for railway-wagon work. It is rarely used at the present day by the cabinet-maker, Russian, Austrian and American varieties, owing to their better working qualities, having quite superseded this native material.

What is known as brown oak is, however, a choice wood for the cabinet trade. The cause of this colouration is usually ascribed to the incipient decay of the trees, and the greater the age of such timber the richer the colour; there are, however, differences of opinion as to the cause of this production. The brown colour is frequently found in young, straight-grown, maiden trees. In many cases trees get pollarded, and in course of years develop burrs and excrescences from which the finest wood is generally obtained. These brown oak trees are not common, and owing to their value are much sought after. One instance, however, where the wood had evidently not been valued, came under the writer's notice some years ago: a good portion of a fair-sized park on the borders of Staffordshire and Derbyshire, having been fenced with cleft oak, remarkably rich in colour and figure and which would probably have been worth in the tree from five to six shillings per foot cube. These trees are mostly found in the Southern and more often in the Midland Counties, but are rare in the North, and only one instance of their having been found in Scotland appears to be known. Whether this beautiful wood is indigenous to Great Britain or not is an interesting question, but so far as is known it has never been found in other countries.

It is generally sawn into veneers, and high prices are obtainable when they are of good description,—the wood making remarkably handsome furniture. A fashion for its use obtained in the United States some twenty years ago, and some fairly large shipments left England for New York.

Stettin–Dantzig Oak.—Oak timber from these Prussian ports was the first that arrived in England to augment the failing supply of her native wood. Principally extracted from Polish forests this timber was formerly largely shipped in the form of planks and partly hewn logs. At the present time the small amount imported is practically all of the latter manufacture. The wood, although not equal in strength or toughness, and deficient in durability as compared with that grown in England, was largely used for naval purposes at the outset and continued in demand for shipbuilding up to about 1865, when iron began to take its place. Large amounts of the wood were however still shipped, its economical use in comparison with native wood for the building of railway rolling-stock being fully appreciated, until wood of the American variety began to supplant it for this purpose. The wood is somewhat yellow in colour, straight-grained, and has easier working qualities than the English, and

A RUSSIAN WAINSCOT OAK LOG

the few shipments that arrive in the markets at the present time are principally used for boat-building, wagon construction, and builders' purposes.

From Memel, another Prussian port, and from Riga and Libau, further oak of Russian and other extraction is shipped. Practically all this timber is exported in the shape of billets or wains ; cot logs, as they are termed. They are prepared by the squaring of two opposite sides of the fallen trunk, the tree being afterwards split through these squared sides right down the centre or heart of the tree. These logs, when shipped and on reaching the consumers' hands, are converted into planks and boards, the wood constituting what is known as wainscot oak. A certain proportion of the boards and planks which are sawn from these billets are more or less covered with the silver grain or figure, which is so much esteemed. A greater quantity of this figured wood is, however, obtained by the American system of conversion, of which an account is given in the pages devoted to particulars of that variety of wood.

These importations from Russian ports are of generally uniform colour, clean, and contain fine figure when properly sawn, and the wood is practically all used by builders for bank, office, and other superior work, as well as by the cabinet trade for the manufacture of the better-class type of fitments, dining-room and library furniture.

The finest wood is of Riga shipment, although formerly the best came from Memel. Of late years, however, Libau has taken the lead in the matter of amount shipped, but the general quality has been decidedly inferior. Shipments are also made from the Russian port of Odessa, in the Black Sea, and much fine wood has been received from this port.

An important trade from these sources is also done in staves for the making of barrels. These are obtained from some of the finest and straightest tree trunks, being cleft or split and afterwards hewn and cut into the various lengths and sizes fitted for the making of barrels of all descriptions. This business is extensive, and a large amount of the finest wood is so consumed.

Austrian Oak.—From the Slavonian and Crotian forests, oak of similarly excellent description to the foregoing is obtained and exported from the Adriatic ports of Fiume and Trieste. It is shipped in the shape of billets, in converted planks and boards, in the form of cleft staves and other ways, the chief markets being France, England and Belgium. It is held in great esteem in England for all purposes where high class oak is necessary, and by many is considered to be a superior wood to the Russian. The export, however, is not so extensive in amount as that of the northern wood, the supply showing some signs of exhaustion.

American White and Red Oak.—Very little of the latter mentioned variety is received in Great Britain, the wood being considered coarse, lacking in strength, and by the cabinet-makers as incapable of being " fumed." The first-named description was originally exported in logs, but this method was gradually superseded by the innovation of converting the same before shipment, and for the last thirty years this system has been generally common, the trade having been a progressively expansive one.

The wood, while not so strong or so durable as English-grown oak, is generally clean and straight and possesses good working qualities, although, owing to the method of quick seasoning which is the rule in the country of its origin it is rendered hard and flinty to the workman's tools. It is largely used for railway-wagon work, for joiners' requirements, and for sills, floor-boards, and other builders' uses : the cabinet-makers are also free consumers, generally using the cheapest grades for the manufacture of the poorer classes of furniture.

As a rule, the boards and planks are fairly well sawn, and the manufacturers excel in their methods of extracting what is known as quartered and plain wood from the logs. To explain these terms it must be understood that all the silver grain or figure of oak lies on what Botanists call the medullary rays. These are rays emanating from the heart or centre of the log and diverging to the outside. They are to be observed in every tree, noticeably strong in the oak, beech, and other trees, but more or less indistinct in others. In sawing an oak tree, all boards or planks so cut that their face falls on these convergent rays, are figured, those sawn contra to these rays being devoid of figure on the face but having it on their edges.

African Oak.—Formerly large shipments of this wood were made from Sierra Leone to Great Britain. The wood was held in high esteem by the naval authorities, but was probably superseded by the gradual introduction of Teak. It is a hard, dense, weighty wood, brown in colour, and notable for its strength and durability, and is rated second-class in Lloyd's list of shipbuilding timbers. So far as is known there has been no shipment to the London market for many years past.

Japanese Oak.—With this wood, the latest description introduced to the markets of the United Kingdom, the remarks on the various varieties of oak commonly in use in England will be concluded. The Japanese, it is well understood, have been conservators of their well-wooded lands for innumerable years, and at the close of the great war which the country had with Russia, began to exploit some of their forest lands. Fair-sized shipments of their oak were made to very many countries, and gradually their enterprise has resulted in the wood gaining a footing in the markets. The timber is shipped in every way that can further its introduction—in logs, in hewn and sawn timber, for railway sleepers, staves for the cooper's use, converted planks and boards, plain and quarter-sawn for the cabinet and builders' trade. The wood is of uniform good colour, mild and adapted for good work and is, when sent in a manufactured condition, well sawn. It is, however, not quartered in so efficient a way as the American wood, is not particularly well flowered or figured, and on the whole lacks the distinctive character of Russian and Austrian wood.

CHAPTER VI

THE MAHOGANIES OF COMMERCE

As Teak among hardwoods stands in a class by itself as representing the ideal wood for durability and strength, so mahogany holds the premier position in the same varieties of wood as a material for ornamental purposes. There are woods that are rarer, some few that are perhaps more beautiful, but none in the popular estimation which, for general all-round qualities, can compare with this wood. Its popularity has not been confined to recent years, for it was known and in good repute under the name of Cedra or Cedrella to the Spaniards when they first occupied the West Indian Islands and the coast of America. It continued under the above name for a number of years, being mentioned as an excellent wood for canoe or shipbuilding by Captain Dampier in 1681. Finding its way to England, probably from St. Domingo or some other West Indian Island, it became known as mahogany, being mentioned in 1730 under that name as an excellent wood for furniture. In 1750 Thomas Chippendale, with his beautiful but at times outrageous designs for furniture, put the seal upon its popularity and, with slight relapses, due to the ebb and flow of fashion, it has continued in favour until the present time. The consumption, in Great Britain and other countries, has shown an ever increasing expansion, and, while supplies have given out in some of the older districts, fresh areas have been found, not only in the Western hemisphere but in the Eastern, and the volume of trade which year by year takes place is extensive.

The number of varieties of this wood which are classed in commerce under the name of mahogany is considerable. From the old Puerto Plata wood from the Island of St. Domingo to the Gaboon wood brought from the French Congo on the West Coast of Africa, there is an extensive range which those in close familiarity can generally distinguish by their external appearance, by noting their texture, growth, colour, and other points of variation. Scientific experts, however, appear dubious as to any connection. Mr. Herbert Stone, an authority on the microscopical identification of timbers, in his interesting book entitled *The Timbers of Commerce and their Identification*, expresses his opinion as follows: " The various species of mahogany and so-called cedar are so confusing that I confess to the inability to make any precise statements either as regards their structure or origin. I know of no convincing proof that any of the American kinds met with on the English market are the wood of *Swietenia Mahogoni*, nor that those shipped from Africa are the wood of *Khaya Senegalensis*. These two genera are very nearly allied to Cedrela and Melia, and it is difficult to separate any of the four from the rest by the characters of the wood. After giving the most careful attention to every detail I lean to the view that most, if not all, of the mahogany met with are Cedrelas."

All the mahoganies noted later are recognised in commerce as mahogany, although doubts as to one or two are at times expressed. Before commencing a few observations on the different varieties, a few remarks on the interesting subject of figure and figured logs may be given. One of the attractions of this fine wood for the manufacture of furniture and other uses is the beautiful figure which is occasionally to be met with in the logs ; one may be what is termed " roey,"

another may contain " broken roe," another may be
" mottled " or marked with what is known as " fiddle-
back wood," a further log may be " ocean-waved,"
and yet another may be " plum-figured." Much
skilful observation, gained by long experience, is at
times necessary to detect signs of this figure in uncut
logs, and considerable speculation is indulged in by those
who wish to secure logs that show signs of figure.

These figured logs after being opened are, if found
sufficiently fine, generally sawn into veneer for the use
of the cabinet and other trades, about twelve veneers
generally being cut from an inch thickness of wood,
if sawn by the veneer saw, while about forty or more
may be produced if peeled off by the knife. High
prices are always obtained when these figured logs are
offered for sale, and it may prove interesting to note
several which are perhaps records for a few of the
different varieties.

From facts given in the *Timber Trades' Journal*
of various dates, we find that the record price for a
figured log was that obtained for one of St. Domingo
wood, sold at public auction at Liverpool in the year 1876.
This log made 30s. per ft. in the inch. Another log,
this of Cuban wood, sold in London in 1912, takes second
place, the log realising at public auction a sum of
22s. 6d. per ft. Yet another from the Island of Cuba,
sold in the sale-room in the following year, realised
22s. per ft.

It is also on record that a log, probably from a
Tobasco district, was purchased in the early days of
the mahogany trade for 21s. per ft., the total cost of
the log having been about £1,000. It was cut into
veneers of five to the inch, and sold to Messrs. Broad-
wood, the celebrated pianoforte makers, a sample of
the wood existing at the present time in the Museum

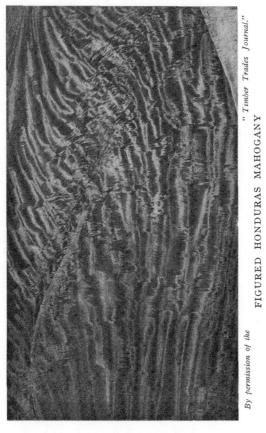

By permission of the "Timber Trades Journal."

FIGURED HONDURAS MAHOGANY

at Kew Gardens. There is an instance of an African log which, selling at Liverpool in 1903 at 12s. per ft., brought to the fortunate shipper a sum of £1,046 5s.; also in the same year, and in the same market, four logs realised 9s. 1d., 9s. 6d., 9s. 5d., and 9s. 6d., per ft. respectively. These latter prices for African wood were, however, eclipsed in 1913, when three logs—all parts of the same trunk—made 13s. 3d., 11s. 7d. and 3s. 6d. per ft. respectively, the total sum obtained for this West African monarch of the woods having been £4,010 12s.

The mahogany arrives from the various ports of shipment mostly in squared logs, in cargoes, or in parcels. Practically all these are directed to London or Liverpool, the latter port receiving the bulk of the West African shipments, the former the greater proportion of the other varieties. With few exceptions all the wood is passed into consumption through the medium of public auctions, brokers at the two ports holding on an average about five sales per month during the year. Very large quantities change hands at these sales, buyers from all parts of the kingdom and also from the Continent being present. These mahogany sales have been carried on for many years past, and interesting recollections of the auctions and how they sold the goods are told by some few old members of the mahogany trade. In London they were generally held, up to about fifty years ago, at Garraway's Coffee House in the City and, not a great many years before this period, at the same place, the goods were announced for sale and sold " by light of candle." A pin was inserted in a candle just below the lighted wick, and the buyer whose bid was made as the pin fell out claimed the lot. This primitive custom of selling goods is still in existence, for, at occasionally held English timber sales that take place in out of the way districts in Cornwall, this old-fashioned method is still observed. It is needless to say that catalogues were not large, probably between 100 and 150 lots. At present-day auctions, and in these times of hustle, close upon, and sometimes over 1,000 are frequent, and a disposal of 200 lots per hour is usual.

CHAPTER VII

SOME ACCOUNT OF THE DIFFERENT VARIETIES
OF MAHOGANY

St. Domingo.—Wood from this Island was probably the first mahogany to reach Great Britain. Shipments were fairly frequent up to about thirty or forty years ago, but since that time the Island has shown signs of exhaustion in regard to the supply, and only small parcels, mostly of inferior and crooked logs, have been shipped.

The wood is of distinctive character, and may be readily identified in the log by its rich and very dark colour, by its texture and, generally, by the peculiar adzing shown in the manufacture of the logs. It is highly esteemed for most work, but the sizes and make of the logs now sent render them, as a rule, only fit for chair-frames and other similar purposes.

Cuba.—This Island, like the last mentioned, a former possession of Spain, has always been noted for the fine mahogany which it produces. It has been a heavily afforested one, in which mahogany has been largely predominant, but with heavy demands on the wood for use in Great Britain, in America, and on the Continent, and in conjunction with the clearances of forest areas for sugar and tobacco planting, there are signs that, although shipments of late years have been fairly plentiful, this source of supply cannot be looked upon as one of long duration. The wood is highly esteemed, and is generally considered the ideal one for the manufacture of furniture, being of hard texture, fine colour, and with a finished surface which takes a remarkably fine polish. Frequently figured, it always has an attractive and individual character, either when used for cabinet-making, in joinery and fitment work, or for any other purpose. As remarked, symptoms of a gradual exhaustion of the wood are not wanting, one sign being the increased amount of small and crooked wood which is shipped to the various markets, the only use for which is chair-making. In most of the cargoes this class of wood forms a large proportion, logs of 12 in. to 16 in. in width being, in any quantity, conspicuously absent. Most of the wood at the present time is shipped from Santa Cruz, but a proportion also arrives from Jucaro, and, occasionally, parcels are received from Santiago and Manzanillo. Mention may also be made of the cedar which is abundant in the Island. America takes perhaps the largest share of this export, but Continental countries are also large consumers ; no quantity of any importance is directed to Great Britain. This wood is practically all used for cigar box-making.

Mexican.—From this Central American Republic, mahogany, varying in description according to the districts from which it is extracted, is exported in fairly large quantities. Wood from the Tobasco district was one of the first mahoganies recognised in commerce, and in later times plentiful shipments from the Minatitlan and Tecolutla portions of the State were shipped. Most of the supplies from the first-mentioned place have of late years been directed to French and other Continental markets, the quantities arriving in Great Britain being quite negligible. Little also of the further descriptions is now sent to English markets, the plentiful supplies formerly sent having been discontinued some twenty-five years. Practically all the wood extracted from this Minatitlan and adjacent districts at the present time finds its way to the nearer and more convenient markets of the United States.

The Tobasco wood has always been held in high estimation, especially by French consumers. The logs exported are generally of good dimensions, sometimes exceeding 36 in. square, well manufactured, and more or less straight : the wood is firm in texture and of good, rich colour, and is fitted for most purposes where length, strength, and the further noted qualities are needed. The quality of the shipments of late years has not however been so good as formerly, a softer, more porous wood having taken the place of the crisp, firm, and harder wood which came earlier.

The Minatitlan and Tecolutla wood, common in the English market twenty-five or thirty years ago, was not comparable with the Tobasco in texture and firmness of grain, but was otherwise satisfactory. This wood, however, is now mostly directed to American markets.

A certain proportion of Cedar logs are generally comprised in the shipments of mahogany that arrive. For this wood there is a general good demand, such logs as are long, straight, and of even grain being much sought after for boat-building purposes, inferior logs selling for cigar box and other work.

Honduras.—The fine wood from this British possession was known, it is believed, as early as the year 1724. It was not until about the middle of the last century, however, that shipments of importance were regularly directed to the United Kingdom. Since then they have largely increased, and during the last fifty years have been one of the leading descriptions used by consumers.

This mahogany is practically all shipped to the London market, and for all-round qualities, whether for the cabinet trade, for builders' use, for the cutting of panels for railway or coach-builders, or for the obtaining of lengths for boat-builders and naval work, there are few other descriptions that are in any way equal. The logs are well manufactured at port of shipment, ranging up to long lengths and of large diameters; the characteristics of the wood are of a good colour, a firm but not excessively hard texture, and an abundance of variety in its growth that fits it for most purposes.

It was formerly, and is occasionally at the present time, known as Bay-wood in Liverpool and the north of England, and was treated some forty or more years ago as a sort of inferior mahogany as compared with Tobasco and St. Domingo, which were then more popular. The name probably had its origin from the fact that the wood was, and still is, exported from the Bay of Belize.

From the adjoining Republic of Honduras, mahogany was formerly shipped in large quantities, especially from the Truxillo district. Not so satisfactory in all-round qualities as that from British Honduras, it yet had good features and was much esteemed, when supplies were regularly shipped, for its large dimensions, its good colour, and straight, mild growth. None from this district, and comparatively little from any other, has supplemented the English mahogany supplies for many years past.

Guatemalan.—The mahogany obtained from this central American State has many outstanding qualities which place it almost on an equality with that from British Honduras. It is even better in colour than that wood, has a similar close, firm texture, and is available in good dimensions. It is not, however, so generally sound in condition, and is usually exported in lengths which, suitable for American consumers, are insufficient for the English markets. No regular supplies are received there, the limited amount that is consigned coming forward in parcels at irregular intervals.

Nicaragua.—Another variety which has certain outstanding features to distinguish it from others, even in adjoining States, has its location in this Republican country. An American firm, with concessions from the Government of the State, formerly operated in certain districts on the East Coast, and fairly large shipments of logs in the round and also of converted timber were shipped to the British markets. This concession has now, however, been annulled, and very little wood either from ports on the east or the west side reaches England.

The timber is obtainable in large dimensions, and the wood, besides being of good colour and of firm texture, is notable for its straight and even grain. It is, however, very brittle, and the wood, generally, has not the individuality about it that Honduras, Guatemalan, and other varieties possess.

Costa Rica.—A much appreciated mahogany is exported in limited amounts from this country, the wood being fairly hard and of good texture and colour. The timber is always shipped in a manufactured form, the squared logs which arrive in European markets being of moderate widths, rather short lengths, and generally in fairly sound condition. Shipments are made in moderate-sized parcels, but the supply on the whole is not large. Cedar of somewhat similar make and description is also forwarded, but in still less quantities than the mahogany.

Panama.—An excellent mahogany is derived from this now famous isthmus. Intermittent shipments have been sent to Great Britain for many years, but no systematic exploitation of the mahogany or other woods of the country has yet taken place. The mahogany is of very straight, even growth, firm and close in texture, and of uniform good colour. It has always been appreciated, but no system of manufacture, such as that practised in the case of the wood from Honduras, has apparently yet been applied, and the logs, mostly of good dimensions, that have been exported have been almost invariably in unsound and split condition. Mahogany-like woods from the same districts, one called Espave the other Santa Maria, and said to be abundant, have been introduced to British markets, but with indifferent results.

Columbia.—A mahogany from this South American State is also known in the English markets. Two varieties are recognised, one from Cispata, the other from Santa Marta. The timber is extracted in large dimensions, the Cispata wood being generally exported in well-squared logs. The first-named variety is very straight-grained but somewhat soft in texture, deficient in colour, and rather liable to become stainy if not quickly converted. That from the latter district is of equally good dimensions, but somewhat coarse in texture and, both in this feature and in its dark colour, is very much akin to the Crabwood of British Guiana and other West Indian Islands. Cedar is also exported from this State, the wood, however, not being comparable with other Central American descriptions.

CHAPTER VIII

THE MAHOGANY PRODUCTS OF THE WEST COAST
OF AFRICA

A LENGTHY chapter on the truly astonishing development in the consumption of the mahoganies of Western
Africa could be written, but the limits of these pages
must necessarily confine remark to a few salient
features in reference to the many varieties with which
this wonderful coast enriches the markets of the world.

Little more than twenty-five years have passed
since several varieties of these mahoganies were first
introduced ; for some years they were received with
anything but favour. Gradually, however, they established themselves, and a steady and progressive import
took place. The statistics appended will show how
rapid the increase has been, and what an extensive
trade has developed in so short a space of time—the
period being well within the memory of most. The
expansion in the demand has not, however, been
entirely due to an increased use of the wood in the
United Kingdom—although that has been large—but
to the steady appreciation which has developed in the
demand from the United States. A large proportion
of the wood landed and sold in Liverpool and London,
estimated to amount to an average of 70 per cent.,
is bought for reshipment to the States : most of the
figured logs, and a great proportion of those of the
best grade, are so re-exported.

The bulk of the total import is landed in Liverpool, a direct line of steamers—the Elder Dempster
Line—bringing over practically all the wood. A portion

is transhipped to the London market, whose supply is further augmented by other transhipments from Continental ports. Practically all the wood landed in Liverpool and in London likewise is sold by the brokers at public sales ; and to show the extent of these auctions it may be mentioned that in Liverpool it frequently occurs that in two sales, generally held in the same week, considerably over 2,000,000 ft. of this African wood is disposed of.

Coming from huge districts which practically stretch from the mouth of the River Gambia to the outlet of the River Congo, an approximate extent of coast-line measuring between 4,000 and 5,000 miles, it is not surprising that very great variety is found in the woods which are exported from the different districts, and which are marketed as mahogany.

Statistics showing the growth of the import of African mahogany during the last twenty-four years—

IMPORTATION OF WEST AFRICAN MAHOGANY LANDED IN LIVERPOOL DURING THE LAST TWENTY-FOUR YEARS

Year	Tons	Year	Tons
1891	3,207	1903	53,450
2	5,808	4	43,686
3	10,290	5	42,654
4	12,214	6	51,432
5	9,159	7	61,503
6	13,820	8	68,063
7	14,877	9	36,194
8	20,636	1910	46,823
9	33,469	1	47,448
1900	41,453	2	60,873
1	30,377	3	83,062
2	25,765	4	77,377

The above figures relate to wood received at the Port of Liverpool only, and do not include the large amount that is transhipped to London *via* this and various Continental ports, neither do they include shipments made to Glasgow and elsewhere in the United Kingdom.

Speaking of the wood generally, it has not the high qualities which characterise the Central American descriptions. Very few of the varieties have the firm, close texture of the older mahoganies, the grain being longer, the wood much more porous. This latter defect mitigates against a high polish being obtained, and for high-class cabinet work it is, therefore, not in favour. It has, however, many qualities which make it popular, supplies are invariably plentiful, it can be obtained in splendid dimensions, both in regard to lengths and diameters, most of the wood is readily worked, and some of the grades are marketed at a very low price : in addition, a good proportion of the wood is more or less well figured. This latter wood is always in demand, especially for the American markets, and some account of the high prices sometimes realised has been given in a previous chapter.

As before remarked, the approximate limits of the country from which these mahoganies are extracted extends as far north as the River Gambia, but, although the wood is known to exist in this locality, no great exploitation has taken place any further than the French Ivory Coast possession. Most of the principal varieties are drawn from this colony, from ports on the Gold Coast and in Southern Nigeria, and others from shipping points in the French Congo. Following these few introductory remarks, some attempt at describing the varied characteristics of the several varieties may now be made.

Assinee Wood.—From a port bearing this name on the French Ivory Coast, this wood is largely exported. It is shipped in excellent dimensions and is practically all sold in the Liverpool market. The colour of the

wood is good and the texture satisfactory, but it is
not generally in favour otherwise than for the fine
figured wood, which, more than in any other variety, is
found in this timber.

Most of the African mahogany, it may be mentioned,
is liable to be affected by what are known as thunder
shakes or wind brakes, a cross-fracture of the timber
which it is often difficult, and at times impossible, to
detect from the exterior of the log. The Assinee logs
are peculiarly liable to this fatal fault, a defect that
renders a log practically worthless when opened.

Grand Bassam.—This variety, also from the same
coast, not only has of late years been the most abundant,
but has occupied a leading position in the estimation
of buyers. Generally shipped in well-made logs of
good dimensions—many being over 48 in. in diameter—
the wood is of firm texture, good, bright colour, and has
excellent working qualities which are appreciated
when the material is used for good-class joinery work.

Sassandra.—A limited amount of the wood from
this district, also in the French Colony, is shipped from
the port of the same name. Notwithstanding the
comparative nearness of this district to the area in
which Grand Bassam and other varieties are obtained,
the wood of this Sassandra variety is totally dissimilar,
being fairly hard in texture and of brownish-red colour.
The logs sent have been generally of good dimensions
and the wood has been received with favour.

From Grand Lahou, shipments of wood of somewhat
similar description to that of Assinee are made in fairly
large quantities, and further supplies are also forthcoming from Half Assinee, Pontadoon, and other places
in the same Colony.

Axim.—From this port on the Gold Coast large
supplies of mahogany are exported, perhaps equal in

amount to the exports from Grand Bassam. The chief proportion is directed to Liverpool and London, smaller quantities to the Continent and to the United States. The wood is of good colour and, generally, of a mild and straight-grained texture. Upon the whole, the make of the logs is not so good as is that of other varieties, and a large proportion of defective logs have been shipped of late years that has not added to the reputation of the wood. This, however, is probably a temporary fault, incidental to bad floating seasons.

Sekondi.—A neighbouring port to the last mentioned. Large supplies are shipped from this centre, the wood being somewhat similar but rather softer than the Axim wood, and perhaps more porous. It is, however, better manufactured, is shipped in logs of large dimensions, and has a good sale.

Lagos. — Shipped from Southern Nigeria, this mahogany, as formerly exported, was held in general estimation as the finest in all-round qualities of any of the West Coast varieties. In the plentiful shipments that arrived some years ago the logs were very unevenly and badly squared, but the wood was of excellent description and compared somewhat favourably with several central American varieties. During the last ten years or so shipments have fallen away and the logs, although rather better in manufacture than formerly, have not been equal in quality to those first exported. It is a distinctly favourite variety among consumers in Great Britain and also in America.

Benin.—From the same Nigerian Coast this wood is exported, the variety being perhaps in as great favour with consumers as the foregoing. It is exported in exceedingly well-squared logs, principally of large dimensions, and the wood is of straight, firm texture

By permission of the "*Timber Trades Journal.*"

MAHOGANY LOGS READY FOR SHIPMENT, SEKONDI

and of good colour, this rather inclining to a brownish or purplish red. It always finds ready buyers when offered on the market.

Sapele.—This variety, mostly shipped from the same port as the last mentioned, is not, to any extent, in favour in the English markets, neither is it appreciated in the United States. It is, however, in demand on the Continent, most of the supply being shipped to Hamburg and elsewhere. The wood is of a cedar-like character, even having somewhat of a cedar smell; this is one of the varieties arriving from the Coast of Africa on which doubts are often heard as to its classification as a mahogany—or even as a cedar. It is hard in texture, rather more brown than red in colour, and has often a straight, stripy roe, which the German consumers favour for cross-banding veneer work in their piano case and cabinet-making. The logs exported range up to very large dimensions and are well squared, but are subject to bad ring-shakes.

Gaboon.—A distinct variety bearing this name is shipped from the French Congo, principally from the port of Libreville. It is of fairly soft and rather coarse texture, light in weight, light in colour—somewhat inclining to brown; has sometimes a strong, coarse, roey figure and, occasionally, one resembling figured birch. It works well as a rule but is sometimes exceedingly strong, and the wood is found at times, owing to its spongy or corky nature, quite to defy the tools of the workman and even the cutters of a machine. This is another so-called mahogany on which doubts are cast as to its identification, and the progression in the growth of its use during the last ten or fifteen years has been most interesting.

The German consumer was first in realising its use, finding in the wood a cheap and efficient substitute for the Cuba and other cedar which had been used for cigar box-making, those central American woods having become not only less abundant but more expensive. It was also found suitable as a substitute for other purposes, and a large trade developed, the wood being known on the Continent under its French name of *Okumé*.

Trial shipments were made to the English markets, but for some considerable time, although offered at low rates, it failed to obtain a footing. Its opportunity came when a shortage occurred in the receipts of other descriptions of mahogany with a consequent rise in values. Consumers then turned to this low-priced Gaboon wood, and a rapid increase in the demand ensued, and how this increase has expanded in very many countries may be seen from the few statistics appended in relation to this timber. It is a useful wood, but this expansion in the demand cannot be put down to any great qualities which it possesses, but rather to the low rates at which it can be placed upon the market. A large proportion is taken by the cabinet trades, but other industries also consume great quantities, notably the ply-wood trade.

STATISTICS SHOWING THE GROWTH OF THE IMPORT OF GABOON MAHOGANY INTO THE UNITED KINGDOM

	1909	1910	1911	1912	1913	1914
	Tons	Tons	Tons	Tons	Tons	Tons
London	441	2,991	5,187	4,791	9,184	8,348
Liverpool	1,818	5,182	4,205	6,946	3,974	2,266
Glasgow	4,529	2,240	5,306	6,214	5,424	1,370
Other ports			70	228		
	6,788	10,413	14,768	18,179	18,582	11,984

Cape Lopez.—Another variety from ports in the French Congo Territory remains to be noticed. It is of generally good description, varying in quality, however, as might be expected, from a wood that is obtained from different localities in so large a district. The best is of good colour, mild, and straight in texture and of fine grain—almost the equal of good Panama wood. The great defect, however, of the logs that are shipped is the worm-holes that are occasionally to be met with, these not being the incursions of small worms on the outside of the logs but borings of sea—probably the teredo worm—which works right through the log. Possibly, better means will soon be found by shippers to prevent this damage, which is doubtless caused by the logs remaining too long on the sea-board. They are exported in fairly well manufactured condition, although somewhat short in length, and the wood holds a favourable position in the estimation of buyers, its use being general in the cabinet and building trades and in other industries where mahogany is used.

African Walnut.—Shipped among the mahogany from most of the districts a wood is found which is sold under this name. It is of varied shades of light-brown and has, perhaps in consequence of its colour, become known as walnut. Putting on one side this colour, it certainly bears every outward characteristic of a mahogany, and this assumption has received confirmation, Professor Boulger, in an appendix to a later edition of his book, remarking, after an examination of samples of this wood exhibited at a Tropical Produce Exhibition, held in Liverpool in 1907 that : " A Benin variety, with the native name· of Apobo Enwina, was a species of *Trichilia* (natural order *Meliacae*) and, therefore, in reality a mahogany, but sold under the name of African Walnut." He describes the sample as of brown to dark-brown colour, having numerous dark veins but no figure.

Another Southern Nigerian sample, however, of

dark-brown colour and rather coarse grain, and with the native name of Owowe attached, he botanically identifies as *Albizzia sp.* (natural order *Leguminosae*), describing it as similar to East India Walnut.   Whether the first-named wood is a mahogany or not, it is impossible—after the wood is stained a mahogany colour and polished—for the keenest expert to distinguish it from an ordinary piece of African mahogany.

CHAPTER IX

THE EUROPEAN VARIETIES OF HARDWOOD TIMBER
COMMONLY IN USE

Walnut.—Three or four European varieties of this broad-leaf, or hardwood timber, are known in English and other markets, namely, that grown in England, and three more exported from France, Italy, and the Black Sea.

The timber is produced from a tree which is said to be indigenous to Southern Europe, and was introduced into England at an unknown date, being formerly much more plentiful than at the present time. It is supposed to be deficient in the power to reproduce itself from seed in Northern latitudes and, probably, this is one cause for the decrease in the supply. At the present time it is to be seen growing much more freely in the Southern and Midland portions of the Kingdom than in the Northern.

It attains to a lofty height, reaching, probably, 70 ft. or over, is of handsome growth with wide-spreading branches, and, in mature specimens, reaches a diameter at the butt of 36 in. or more. These large boles or butts are more esteemed than those of lesser growth, the wood gaining by age not only a closer texture but also more heart-wood with good colour and dark stripy markings. Such wood is usually found in the butt and principally at the base, even the rooty portions often containing fine figured pieces. This walnut was, formerly, largely used for gun-stocks, but at the present day, owing principally to its scarcity, is only used for high-class sporting rifles. When more plentiful, it was largely used by the cabinet trade, most

74

of the furniture at the end of the seventeenth and commencement of the eighteenth century, before the mahogany period set in, being made of this wood. There is an active demand in present times for the timber, and it realises good prices, but it is scarce, and little beyond isolated specimens come into the market.

French Walnut.—This wood is of similar character to the English, but, while generally well-coloured, has not perhaps so much stripy figure as the English-grown, at any rate that is the case in the wood exported, which consists of planks mostly cut from the upper part of the trunk.

Italian.—This perhaps shares, with the wood obtained from Circassia, the highest appreciation of consumers. A fashion prevailed about thirty-five years ago for this wood, great quantities being in demand for chair-frames and other work in the cabinet trades, and to fill this demand large exports of planks from Genoa and other Italian ports were received. They were mostly of narrow widths and contained much sap, but found a ready market, the wood being sold by weight. The fashion abated after some years, and the use of the wood was supplanted by American walnut and other woods. Occasionally, parcels of better-class planks have since arrived at intervals, such timber selling readily, when of prime quality, at good prices. The wood is of fine character, and when of stripy description makes a handsome material for high-class furniture ; it has, however, some liability to the attack of small worms, under which the wood slowly crumbles away.

Circassian.—Another fine wood is shipped from Batoum and other ports in the Black Sea, the wood being frequently known as Black Sea Walnut. It is extracted from the hillsides of the Circassian Mountains, and is shipped in hewn logs some 6 ft. to 10 ft. long, which are cut down close to the root to obtain the finest wood. These logs are sold in the English market by weight, and a good proportion are cut into veneers for use in the pianoforte trade. There is also a large re-export trade transacted with buyers in America, the wood being, at the moment, in favour in that country. The burr walnut, that was so much in fashion about forty years ago, was also obtained from Circassian sources. French merchants were largely interested in this trade, and, in addition to parcels of burrs that came direct to these markets, others were shipped to Marseilles, from whence they were sent to Paris, sawn into veneers, and mostly sold to English buyers. Very high prices have been realised at times, and little wood otherwise was used by the pianoforte trade for veneering cases, a good proportion also being consumed by the cabinet trades. The wood in this burr form is quite out of date at the present day and is rarely seen.

Elm.—Several varieties of this tree are common to most European countries ; two only will be mentioned, however, these, of native but indigenous growth, being known as Common Elm and Wych Elm. The first-named is too well known to need a lengthy description. Grown in hedgerows or in avenues, it is common in the Southern Counties of England, but is, perhaps, seen in more perfection in districts in Northamptonshire, Warwickshire and Worcestershire, where its picturesque growth attains to a height of 70 or more feet, with a girth of 7 or 8 ft. In its growth, the trunk has a habit of forming pockets where a branch has been lopped or broken off, this cavity afterwards becoming surrounded by fresh growth of the wood. Curious discoveries are sometimes made when the trees are felled and opened, stores of acorns deposited by squirrels, stones, or other things, and perfect bird's-nests with the remains of eggs being sometimes found. This latter curiosity has been seen by the writer embedded in a hole with an encasement of about two inches of wood grown after the nest was deposited.

The wood obtained from these trees is of light-brown colour, much twisted in the grain, fairly hard, tough and porous. There is usually a good margin of clearly defined sap-wood which, however, is as durable as the heart-wood. It is difficult to split, and is exceedingly durable when placed in a constantly damp position, and equally so if kept in a perfectly dry one, but will not stand an alternate treatment. As an instance of its durability, an example is afforded by the conduits formerly employed by the now defunct New River Co. These are often found buried in the London Streets, simply trunks of timber bored from end to end, and spigotted one end into another. These are brought to light practically in as sound condition as when they were placed there, perhaps a hundred years ago. The wood, however, notwithstanding this virtue of durability, has not a great many users, it being inclined to twist and warp and, as remarked, is difficult to work. It is largely used for packing-case ends, for the bottoms and sides of wheelbarrows, for ships' blocks owing to its smooth wear, and especially for the making of coffins.

Wych Elm.—This is a variety that is supposed to be indigenous to Great Britain. It is rarely seen in the Southern Counties, but further North and in Scotland it is common, being in fact sometimes known as Scotch Elm. It rarely, but occasionally, grows to the height of common Elm, and has a rather pendulous and more graceful habit in the disposition of its branches. The wood is of lighter and more uniform brown colour than the common variety, and is also much straighter

and less contorted in the grain than the better known
kind. It has good working qualities and, in addition
to its durability, has considerable longitudinal strength
and much elasticity after steaming. The wood lends
itself to bent work, and is esteemed by wheelwrights for
hubs, felloes, and other work.

It has in many respects some resemblance to the
Rock Elm of the Western States and Canada, on which
remarks will be made in a further chapter dealing with
the woods of the Western hemisphere.

Lime.—A well-known tree, that is common in all
parts of Europe. Formerly the favourite for planting
in the London streets and parks, the Lime has of late
years been supplanted by the Eastern Plane. The
tree produces a soft, yellowish-white wood which,
lacking durability, is of no great value. It is, however,
although soft in texture, very close-grained and, therefore, esteemed as a material on which leather may be
cut, being in considerable use for this purpose by
saddlers and other leather-workers. For use in the
manufacture of pianoforte actions it is also in demand,
principally for the foundations on which the ivory
or celluloid facings are attached for the keys. It has
besides always enjoyed a reputation as an efficient
material for wood-carving, the work of Grinling Gibbons,
in St. Paul's Cathedral and elsewhere, having been
executed in this wood.

Ash.—This is a very widely distributed tree, being
common to most parts of Europe and other Continents.
Grown, more or less, in most parts of Great Britain, the
tree produces perhaps an even more valuable wood at
the present time than the oak. No variety from
other countries can compare with it when well grown,
either in toughness, strength, or other qualities, and
it is regrettable that more endeavours are not made to

supplement the supply by judicious planting. The tree is well known, the best localities for its growth being moist valleys, where, planted in coppices as it should be grown, it develops straight, clean trunks, which are so much sought after. That grown in isolated positions, in hedgerows, produces timber of far less value. The wood is of greyish-white colour, close in grain, tough and elastic. It has considerable, but not clearly defined, sap-wood, which, however, is quite as usable as the heart-wood. There is a tendency to turn black in the centre when old, this feature being perhaps an indication of approaching decay. The tree has the habit of sprouting from the stools or roots after being cut down, and this second growth in some American varieties is supposed to be superior to the first. In certain districts of England the method is used for obtaining a supply of young shoots which may be used for crate, stick, and other purposes.

The uses for the wood are manifold ; chiefly, however, it is employed in the construction of coaches, carts, motor-car bodies and other vehicles, and for general wheelwrights' use. It is also largely used in the manufacture of requisites for sports, for tool handles, for butchers' blocks and an infinite number of other purposes. The supply of British-grown timber is never equal to the demand, and has to be supplemented by wood from other sources, a large import arriving from America in the shape of unmanufactured butts and converted planks and boards, a fair amount from Russia, some little from the Hungarian forests, and from Japan.

Plane.—Known as the Eastern Plane, this tree was introduced into England. In the Southern Counties it grows freely, but is practically unknown in the Midlands and further North. It is a handsome tree when allowed free development, and is well known in London where it appears to thrive on the soot-laden atmosphere. The wood is of little value ; it is smooth in grain, light-brown in colour, but not durable. When cut on the medullary rays it shows a pretty beech-like figure to which the Americans, when exporting their variety of Western Plane, attach the fancy name of Lacewood. Shipments of this Western Plane were made from the United States some years ago to Great Britain, in the shape of converted boards. Being a rather attractive wood it was tried by the cabinet trade, but, owing to its liability to twist and warp and its bad working qualities, had but a short season of success.

Poplar.—Three species of this tree are fairly plentiful in Great Britain and other European countries—the Black, Lombardy, and Aspen. They produce a grey-white, soft, but tenacious wood, which, durable enough when kept dry, is not, however, held in any great esteem. Its chief uses are for cart-bottoms and similar purposes, where its non-splitting qualities are appreciated. Shipments of an American species have been occasionally received in the markets of the United Kingdom under the name of Cotton-wood, but with little encouragement ; Germany, however, it is understood, consumes fairly large quantities.

Beech.—Said to be a native of Europe, but is not only abundant in this Continent but plentiful in Asia and Africa. It is of free growth in the United Kingdom, particularly favouring the chalk districts of Buckinghamshire, where the beech-woods are famous. It is one of the most attractive trees of the woodlands, with its lofty growth, sweeping branches, and beautiful spring and autumn-tinted foliage. The beech produces a whitish-brown wood, with a tinge of pink which appears after it is sawn. It has a silky grain, cleaves well, works satisfactorily, and has a pretty but small figure when the wood is quartered. It is durable under water or when kept dry, but not under varied conditions. Its uses are many, the first and foremost being for the manufacture of common chairs, and also for parts of the frames of better-class work. Being a hard, close-grained wood, it is largely used for turnery purposes, and is the material employed for what are known as wrest planks in the making of pianos. It is also largely used for the making of plane blocks and the handles of tools, besides an infinite number of other purposes which it is unnecessary to mention. The native supply is about equal to the demand, but imported wood from Germany, in the shape of sawn planks, is received in small parcels from time to time.

Alder.—A small low-growing tree, common throughout Europe. It is generally found in low-lying, swampy places, seldom exceeding 40 ft. in height and of small diameter. The wood is of whitish-brown colour, changing to a redder shade on exposure to the air. It has a very fine and even grain, and is exceedingly durable when employed in water, but will not stand in alternate positions. It is not greatly used and is of comparatively little value, the chief uses being the making of clog soles, brush backs, and for turnery purposes.

Hornbeam.—This tree is supposed to be indigenous to Great Britain. Like the foregoing, it is one that does not attain to a great size, and is moreover not abundant. The wood is yellowish-white in colour, fairly heavy, extremely hard and close grained, and exceptionally strong when subjected to vertical pressure. It is an admirable material for turnery purposes and the making of wooden screws. It is largely used by engineers for cog-wheels, and a fair amount is consumed for the making of parts in pianoforte actions, as well as by boot-tree manufacturers. The small native supply is generally sufficient, but it is occasionally augmented by small parcels of French wood, which are shipped in the form of planks.

Sycamore.—This is a species of Maple which is found in most parts of Europe, being fairly common in Great Britain. The tree, which generally grows isolated, attains to a moderate size and is not so abundant as might be desired. The wood is generally of fine white colour, especially that obtained from young trees, that from those of older growth being often of a brown tinge in the middle. It has a fine, close, and even texture and, occasionally, a handsome mottled figure, this latter wood being much used by makers of violin and other musical instruments. It is an admirable wood for the turner, and is much used for rollers of calico printing and wringing machines, also for bobbins and many other similar purposes. The supply of sizable timber is never in excess of the demand, and good prices have to be paid for the wood.

Willow.—From the osier and lower forms, many varieties of willow are abundant. Few, however, have any great commercial value, their growth being perhaps more useful for the retention of river banks than for their timber. The wood is somewhat similar to that of the Poplars, being soft, light in weight, and not splitting or splintering readily. It was formerly employed in the making of buffers of railway trucks, and is sometimes used for cart bottoms, but is chiefly esteemed for the cutting of shoots from the pollarded trees for making crates, hurdles, baskets, and other similar work. For the manufacture of cricket bats it is renowned, one species, the Huntingdon willow, being preferred for this purpose and realising high prices when of good growth.

By permission of the "*Timber Trades Journal.*"

THINNING HORNBEAM WOODS, EPPING FOREST

Many osiers are cut down from beds in low-lying river districts ; the supply of these, however, not being sufficient to fill the demand, imports from Holland in large quantities take place

Horse Chestnut.—This is a tree producing timber of no commercial value. It grows to fine dimensions, with dense foliage, and is notable for its handsome appearance, the avenues of these trees in Bushey Park, which form such an attraction when in bloom in the spring, being an example of their natural beauty. The wood is white in colour, soft, close and even in the grain, but twists and warps badly, and is altogether deficient in durable qualities.

Sweet or Spanish Chestnut.—This is found in abundance in Southern Europe, but is thinly scattered throughout England. It attains to very large dimensions, and is a stately and beautiful object when growing. The wood, however, is not in any great use. It is brown in colour, moderately heavy and hard, and has a straight, even and porous grain. It has a decided oak-like figure when cut on the medullary rays, and is sometimes used as a substitute for that wood. Very durable when submerged in water, it is consequently used for sluices, flood-gates, and similar work, It will not, however, stand in alternate wet and dry situations.

Olive Wood.—A tree that is well known in Mediterranean regions, and which grows to a medium size. It produces timber, from its contorted and twisted trunk, which is of no great use except for the making of fancy-turned articles and small pieces of cabinet ware. The wood obtained is of yellowish-brown colour, close-grained, and hard in texture. It has, usually, a fine stripy figure in its twisted growth, and its handsome appearance renders it fit for the above-named purpose.

Box Wood.—This is obtained from a tree of shrubby growth which is indigenous to a portion of Europe and parts of Asia. It is a slow growing, low, and bush-like tree that is occasionally to be met with in England, the timber produced being small and of comparatively little value : it may be frequently seen in an extremely low form, in many cottage and other gardens, where it is grown and trimmed for borders. In parts of Turkey and Persia, however, it grows more freely and attains a larger size, and it is from these localities, shipped from ports in the Black Sea, that we obtain the most appreciated qualities.

The wood is exported in small billets or sticks, generally about 2 to 3 ft. in length, and from about $1\frac{1}{2}$ in. to 6 in. in diameter, most of the parcels being chiefly shipped to Liverpool, from whence it is distributed.

Light yellow in colour, it is extremely hard and close-grained, having a dry weight of something like $53\frac{3}{4}$ lbs. to the cubic foot. It works and turns well, and has a silky lustre when finished from the tool.

Formerly, the chief demand for this timber emanated from the wood-engraving trade, the material having been used for the cutting of blocks for illustrative purposes since the fifteenth century. No other wood could be found with the same close, fine texture, so well adapted for the purpose, and high prices were for many years obtained. Of late years, however, the camera and other processes of obtaining illustrations have superseded this older method, and the consumption for this purpose has declined. The wood has still many miscellaneous uses, being largely employed in the manufacture of rules of all kinds, for turnery work of all descriptions, for the making of tools and tool handles, inlaying and other work.

A further variety which is, however, not held in

such high esteem, is shipped from Central America and Brazil. This wood is imported in straight, round logs, up to about 10 in. or more in diameter, is of greyish-yellow colour, perfectly straight in the grain, but has not the dense and horny character of the European varieties. Two other species are also exported from South Africa—Knysna and East London—both, like the above West Indian variety, being inferior to the Turkey or Persian wood.

Many additional woods are in use, such as Holly, Yew, Pear, and a few others, but so limited is their use that, except for this slight allusion, it is unnecessary to add further remarks.

CHAPTER X

THE great East Indian Empire is richly afforested with fine timbers of almost innumerable varieties, about which exhaustive information is available in the publications of the Indian Forest Department. Most of the forests are under the control of this Department, the woods being exploited under their direction, schemes for systematic reproduction being formulated under their authority, and in addition they issue abundant information as above stated, covering the whole subject.   There is apparently a great native consumption of these Indian woods, for out of an approximate total of 5,000 woody species which are said to exist in the country, and of which rather more than half are timber trees, not more than about a dozen are regularly supplied to the outside markets, and only one, namely, Teak, is shipped in any considerable quantity.

Satin Wood.—The introduction of this fine wood probably took place for the first time about thirty or thirty-five years ago, when logs and more often planks, sawn into boards, were brought to the English market. Most of the wood was finely figured, and was readily sold, notwithstanding the difficulties of cutting these boards into veneers.   Gradually, with the establishment of a regular demand, nothing but timber in the round was shipped.   It now arrives in this natural condition, the logs mostly ranging from 10 to 20 ft. in length, and with a diameter of about 16 in. to 24 in.   The wood is of yellow or yellowish-brown colour, very hard, dense, and with a bright lustre.   Many of the logs are finely figured with an attractive mottle, and high prices are secured for these, 5s. and 6s. per foot of 1 ft. by 12 in. not being uncommon.

The wood is largely used for brush backs, but much of the finest is bought for conversion into veneer for cabinet-makers' use. It is a difficult wood to work, however, not taking glue readily, and requiring very skilled workmanship, and is thus only seen in furniture of good grade for which a sufficient price can be realised to repay the cost of manufacture. It will not compare in colour with the fine satin woods of the West Indian Islands and has not, moreover, such bold and striking figure as some of those woods possess, but it is, nevertheless, when finely figured, a most attractive and beautiful wood. It is interesting to note its uses in India, compared with those elsewhere, among the many cited being building work, bridge construction, wharf piles, felloes, and durable as a material for railway sleepers, but too valuable.

Rosewood.—Generally known as East Indian Rosewood, and sometimes in India as Bombay Blackwood. It grows in fair quantities in Central and Southern India and other parts, reaching its greatest size throughout the Western Ghats. It is rather plentifully exported to Great Britain, where it arrives in round logs ranging up to a diameter of about 36 in. or more. The wood is of dark reddish purple, with black marks, and has a hard but rather coarse and open texture. It is a handsome wood and valuable, being employed in limited quantities for cutting into veneers for the use of pianoforte makers and others : the greater proportion, however, is re-exported, a good demand being found for the wood on the Continent.

Padouk or Andaman Redwood.—This is a remarkably handsome wood which is obtained from the Andaman

Islands. A further species is found in Burma, but this is of inferior character. The wood was first put upon the market about twenty-five or thirty years ago, but, although intermittent shipments have from time to time been made, it has never obtained a very permanent footing. A good proportion, practically all, of the parcels received of late years in England have been transhipped to the United States, where there is an occasional demand.

Shipped in squared logs of large dimensions, the wood is of most striking appearance, being of a rich deep red colour, streaked with black, which, however, fades on exposure. It is hard, durable, seasons well, and takes a good polish, but is somewhat cross-grained and difficult to work. It is apparently used in a variety of ways in India, for the frame-work of ploughs, for carving, for the making of sticks, for railway sleepers, and in the construction of casks.

East Indian Walnut.—A large deciduous tree common in the Himalayas from Bhutan westwards, extending into Afghanistan. The wood produced is dark brown and fairly uniform in colour, hard and silky in texture, seasons and polishes well, and is understood to be fairly durable. Unlike other walnut woods, no great consumption is found for the wood in the European markets, but occasional shipments are made. Well-figured logs have also been shipped and have sold readily. It is said that fine burrs are to be obtained from these trees ; none however, so far as is known, have been seen in the English Markets.

Red Cedar, Moulmein Cedar (Toon).—Variously named, another large deciduous tree common in most parts of India has to be noticed. The wood is of bright red colour, is soft, open, and even-grained, is easily worked, but rather brittle. Largely used in India, has not obtained much favour elsewhere, although shipped in well-squared logs of large dimensions. Has no cedar scent, and its too bright colour is perhaps an obstacle to its use for cigar box-making.

Gurjun.—The produce of a large tree, from which logs of excellent sizes are obtained ; in the Andamans, squares of 24 in. diameter and with a length of 60 ft. can, it is understood, be extracted. The wood is of reddish-brown colour, soft to moderately hard in texture, rough, and not of particularly durable character. It is used in India for general ship and boat-building purposes, and is an inferior material for house-building. It was introduced into London as a likely wood for street-paving, its oily nature being urged as a merit for this purpose.

Macassar, Ceylon, and Andaman Marble Wood.— Among the varieties of ebony which are found in India the two first named are regularly exported, and samples of the latter have also been shipped. The Macassar is extracted in moderately long billets of good average size, occasionally ranging up to 24 in. or more in diameter.

The wood of these varieties is of generally hard, close and fine texture, and of a very heavy nature. Contrary to the popular conception of the colour of this wood as jet black, it is usually varied, the Macassar being of mingled shades of brown and black ; the Ceylon, rich dark purple with black streaks ; the Andaman Marble or Zebra wood being very finely striped with alternate but irregular streaks of grey and black. The two first described varieties have a moderate consumption, being principally used for the manufacture of walking sticks, for turnery purposes, and the making of fancy articles of cabinet ware. The Andaman Island wood has scarcely been tried, but is by far the most handsome in appearance of the three descriptions.

Teak.—This valuable wood not only ranks as the most important among the many varieties of fine woods that are grown in India and Siam, but stands, in the estimation of consumers, in a premier position in regard to its use as a material for constructional purposes. It is so classed at Lloyd's, being placed in the list of shipbuilding timbers in a position alone, and above oak and every other wood.

It was a timber held in high estimation in India in the days of the old East India Company, and its good qualities were gradually brought to the notice of the naval authorities in England, by whom it was first used about a century ago.   By degrees its qualifications as a wood for construction began to be appreciated in other directions, and upon the growth of the railway system and, later, of tramway means of locomotion, in this and other countries, it was found to be a material of admirable description for the construction of the vehicles used.   It also found favour among architects, and large quantities, in the aggregate, are at the present time consumed by the building industries.   The consumption has also greatly expanded since its introduction in Asiatic markets nearer to its source of extraction, the bulk of the output of India being shipped to other centres in the same country, and to markets in China and Japan.

The wood shipped from India is practically all obtained from forests in Burma, and from the ports of Moulmein supplies were first distributed. Later, Rangoon in the same State came to the front and eclipsed, in the amount of timber she exported, the older port.   Afterwards supplies of the same description of wood were exported from the neighbouring kingdom of Siam, and, although at first the wood was held in less esteem than the older varieties, gradually established itself in favour, and large and regular shipments from the port of Bangkok followed.

In India the tree from which this fine timber is obtained is found growing in mixed forests in Central and South India and Bombay, but even in far larger quantities in the forests of Burma. As before mentioned, most of the afforested areas in India are under State control, and in the case of Teak rigorous measures are taken, not only in regard to the number and girth of the trees which are to be felled, but also in the perhaps more important question of replanting and establishing a continuity of supplies.

Such trees, for which State authority has been obtained for felling, are first girdled, a ring of the sap right down to the heart-wood being removed from near the base of the trunk. This, effectually performed, arrests the flow of sap and puts an end to the life of the tree. The trunk is then left standing for two years, the wood in that period partially drying, and so being reduced in weight that it will float on the rivers to destinations where it is manufactured into logs for export.

The traction of the great trunks from the forests to the rivers in proximity is a task of great magnitude. Frequently they have to be hauled long distances over rough and practically unmade roads through jungle and forest, the work being both arduous and expensive. Elephants are largely employed in this heavy work, and the sagacity and enormous strength of these animals largely alleviate the difficulties of the task.

On the rivers, swollen by season's rains, the trunks of these trees after being sufficiently reduced in weight to float—which they would not do when green—are brought down to mills adjacent to ports of shipment. Arrived here they are, by means of modern machinery, converted into logs, planks, boards and scantlings to suit the needs of consumers in the various markets for which the wood is destined.

Here again the elephant, the animal with the smallest brain, in comparison to its size, of all existing beasts, is brought into use, his uncanny intelligence and his great strength making him indispensable. Some facts from these localities in regard to the latter qualities possessed by the animal are interesting.

As regards strength, the elephants are reported as being able to lift on their tusks, with comparative ease, a log weighing half a ton, and to drag, when sufficient hold is obtained, a trunk weighing 3 tons. But to realise their power, of which they appear to be proud, one has, it is said, to see them at work in the jungles and creeks when launching the trees into the rivers. Here, sometimes, when elephant and log are buried in soft mud, the animal has tasks which bring out his capabilities, and his qualities are seen to their best advantage. It is related that some years ago a tug of war was tried at Moulmein between an animal belonging to the Forest Department and Sepoys attached to a Madras regiment : it took, it is recorded, 130 men to hold the elephant.

With a mahout, or keeper, on his back he practically does all the heavy labour at the mills, picking up great logs with his trunk, conveying them to the machinery, and piling them afterwards as intelligently and effectually as a pile-stacker in a London dock. They are most regular and methodical in their habits, and in one respect similar to human labourers, it being said that, when the bell proclaiming the time for leaving work is heard, no temptation whatever will induce them to continue, even if they are in the course of piling a log.

Teak is of fairly heavy weight, a cubic foot of seasoned wood having a reputed weight of 45 lbs. It is of uniform golden-brown colour, very occasionally in the Burmese varieties, having black streaks on its surface, and is still more rarely found with some little figure. Now and again it has also a greenish tinge when cut, this colour, however, disappearing on exposure to the air. The grain of the wood is open and slightly coarse, and, when planed or worked, a most unpleasant smell is emitted, which, however, like the green colour above mentioned, disappears with exposure. There is a gritty substance in the pores of the wood which blunts any edged tools, but, upon the whole, it has admirable working qualities. Its strength and durability is unequalled, especially in regard to the latter quality, even in the moist heat of the tropics the wood being practically imperishable. Its great and peculiar virtue, however, and that which first brought it into notice, in addition to its many other good qualities, is its possession of some oil or substance in its constitution that prevents the rusting and corrosion of iron when that metal is placed in contact with it, and, for this quality alone, it is highly esteemed in naval circles all over the world, no wood having yet been found that possesses similar features.

The timber can be obtained in excellent dimensions, both in regard to lengths and sizes. Formerly, all the wood received in England was in the form of squared timber, but, within the last thirty or forty years, much besides has been sent in the form of planks, boards, scantlings and other conversions. Almost since the naval authorities put their seal on its qualities, the finest logs in length, diameter and quality, have been reserved by shippers for export, the selection being known as " Europe quality."

As before observed, while a great proportion of the supply is utilized for shipbuilders' work, the consumption for other purposes has grown considerably and, notwithstanding the efforts of the Indian Forest Department, the supply from these well-guarded and systematically arranged forests has rarely exceeded the demand, and the wood has shown, more particularly of late years, an almost continuous appreciation in value. Prices have reached such a point during the last few years, especially for the higher grades of the wood, that its uses have been somewhat restricted where it has not been an absolute necessity. Many substitutes have been of late tried, but, although several with excellent qualities have been put upon the market, none have approached Teak in its all-round good qualities, particularly in reference to its adaptation for use with iron.

Two or three varieties of wood are at the present time being offered as substitutes: In or Eng wood, a product of Burma; Yang, another species of Eng, exported from Siam; and Krabark, a wood shipped from the latter country, but at present unidentified. These, all useful woods, are being slowly established, and, to a small extent, take the place of Teak in its minor uses.

Serayah.—An unidentified wood imported in the form of planks from Borneo and the Malay Peninsula. It is of uniform reddish colour, open grained, and strong, and, though inclined to twist and warp, is fairly workable; it does not, however, take a finish, being coarse and occasionally of woolly texture.

Amboyna—A scarce and valuable fancy wood obtained from one of the Molucca Islands bearing that name. The wood is received in the form of a burr, this woody excrescence being produced from an unidentified tree, which is presumed to be of coniferous growth.

These burrs are but rarely met with, and contain beautifully mottled wood, brown in colour. The burrs are cut into veneer and used for fancy cabinet work and for inlaying.

Thuya.—A very similar wood to the preceding and hardly distinguishable. This is also the burr of a coniferous tree which is found in Algiers and Morocco. It has a fragrant odour, is finely mottled, and is used in a similar way and for the same purposes as Amboyna.

CHAPTER XI

MANY varieties of timber, principally of the hardwood division, and mostly fitted for heavy constructional work, are grown in the Australasian Colonies. Most of them belong to the Eucalyptus family, the Continent being richer in this genus than in any other. They are chiefly strong, heavy and durable woods, fitted for the above-mentioned work, and of them about six or more varieties are exported in more or less quantities to the English and other markets. Some few descriptive notes may be given in regard to these, and also to a few further woods which, suitable for more decorative work, are occasionally shipped to Great Britain.

Jarrah.—This timber and Karri are perhaps the most important that are grown in Western Australia. The tree from which it is obtained grows to a height of 90 to 120 ft., the trunks often showing a length of 50 to 60 ft. to the first branch, and being from 3 to 5 ft. in diameter. The wood is of deep red colour, hard and dense in texture, its weight, when fresh, being about 70 lbs. to the cubic foot, and, when seasoned, about 10 lbs. less. It is largely used in England for street-paving purposes, and in Australia, Cape Colony and other countries, for piles, jetties, bridge building and railway sleepers. It is exported, cut to fine dimensions, and is pre-eminent, not only for strength and durability but as being impervious to sea-worms.

Karri.—One of the largest timber trees of Australia,

which frequently attains to a height of 200 or more feet ; its straight symmetrical bole towers upright and overtops all other eucalypts in its vicinity. The wood is, in its main characteristics, similar to the Jarrah, but is understood to be less durable, although, in some respects as a paving material, it is superior.

Ironbark.—This further species of the Eucalyptus family is another timber which holds an important position among the timbers of Australia. There are four varieties, three of them being of especial value, fitted as they are for bridge work, beams, railway sleepers and other similar work. The trees attain a height of 100 to 150 ft., with a diameter of 2 to 5 ft., and the timber, hard, tough and extremely difficult to work, is no less than 73 lbs. to the cubic foot in weight. The wood from these varieties differs in colour, the three most in favour varying from white to grey, the further one being of a reddish colour.

Blackbutt.—A tree growing from 50 to 150 ft. in height, and with a diameter of 2 to 4 ft., is also one of the most valuable and esteemed of the Eucalyptus family, being adapted for similar purposes to the Ironbark. The wood varies somewhat in colour, but is generally of a greyish brown. Most difficult to work either with saw or plane.

Blue Gum.—Another similar wood of durable charac- ter, produced from a fine, shapely tree, whose symmetrical growth attains a considerable height. It is largely used for similar constructional work as the foregoing

Tallow Wood.—Is so called because of its greasy nature. It is a wood that, after Blackbutt, ranks next in general excellence. It is strong and durable, and, having good working qualities, is in favour for building purposes. It is of pale brown colour, difficult to split, and is reported in Australia as being an admirable wood

JARRAH

for street-paving purposes ; very little of this wood has, however, been exported to English markets.

Red, or Forest, Mahogany.—Is a handsome timber which is loosely called a mahogany, the timber being really of the same Eucalyptus family as the Australian woods already noted. A fair quantity of this timber, generally in a converted form, reaches the London market. The wood is of deep brownish-red colour, hard, dense and heavy in weight, has fair working qualities, but the grain is occasionally rather twisted.

Silky Oak.—No true oak is found in the Colonies, but to this timber (*Grevillia Robusta*) the above fancy name is applied. It is a handsome wood of light colour, with a fine oak-like grain, and with blotched oak-like figure. It is largely used by the cabinet trade in the Colonies, and has been introduced into Great Britain, although with no very encouraging results.

Black Bean.—Has some resemblance to very dark walnut, and is a fine-grained wood that takes a high polish. Occasionally it shows fine figure, and is much used in the making of Colonial furniture, but it is said to absorb glue unreadily owing to its greasy nature. This wood, like the preceding, has had an introduction at sundry times to the English markets.

Myall.—Is an intensely hard, heavy and dark-coloured wood with a scent of violets. The tree does not attain to any large size, seldom exceeding 30 or 40 ft. in height, with a diameter of 12 in. to 18 in. It is much used for turning purposes, and was at one time, and is now to a limited extent, used for the making of tobacco pipes. Has been used by the Ordnance Department for the manufacture of spokes for gun carriages, and is a wood of the Acacia species.

Red Cedar.—This is identical with the Toon of East India and other Asiatic countries, of which a short

SILKY OAK

By permission of the   Agent-General for New South Wales.

description has been given in the chapter devoted to Indian woods. It is reported as being plentiful in the Colonies, and is largely used for various purposes. It has not, however, when tried in the United Kingdom, found much favour in competition with old-established varieties from Central America.

From further Asiatic countries other varieties of hardwoods, in more or less limited quantities, are shipped to Great Britain. Japan, in addition to the supplies of oak, an account of which has been given in a previous chapter, sends to the English and other markets consignments of ash. This timber is shipped in squared logs, somewhat short in average lengths, but of good diameter, and the wood, although not comparable with English or American for toughness and strength, has found some favour. It is lighter in weight than the English wood, of straight, clean grain, and possesses good working qualities. It is perhaps more fitted for furniture making than for the uses to which ash timber is generally put, and, should a recurrence of a fashion for this wood, as a material for furniture, again occur, would probably be in great demand.

From Borneo several woods are exported, the principal being a cedar, which, under the name of Borneo Cedar, has a moderate sale in the English markets. In addition, camphor wood is shipped, and one or two further varieties.

From the Dutch possessions of Java a regular export of Teak wood takes place. This variety, under the name of Java Teak, is a regular article of commerce both in the English and Continental markets. It is not comparable, however, with the timbers of Burma and Siam, and realises very inferior prices compared with those made for the superior kinds. It is generally exported in short logs and flitches, mostly of hewn manufacture.

HONEYSUCKLE

By permission of the   Agent-General for New South Wales.

BLACK BEAN

The arboreal wealth of the Philippine Islands is, under the restless activity of American exporters, being rapidly developed. A large number of species—about 2,500 it is said—have been discovered, and many, of more or less value, have been marketed in the States. One or two woods in a converted form have reached the English markets of late. They somewhat resemble a coarse mahogany, but are perhaps more similar to the crab-wood of British Guiana. As yet no definite result in regard to their appreciation by consumers has been attained.

CHAPTER XII

SOME ACCOUNT OF THE HARDWOOD TIMBERS OF CANADA
AND THE UNITED STATES EXPORTED TO
GREAT BRITAIN

THE coniferous timbers of these two Western Countries have already been dealt with in a former chapter, and it now remains to allude to the abundant wealth of the further hardwood section, confining the remarks to such varieties as are commonly met with in present-day commerce of the United Kingdom.

While the supplies of the broad-leaf timbers received from Canada are negligible in quantity, compared with the large shipments of soft woods, those shipped from the United States are abundant in comparison with the lesser quantity of coniferous woods exported.

Ash.—In the United States alone there are said to be no less than sixteen species of this wood, more or less valuable, and in the Canadian Dominions there are also many more. Only two, however, of this number are exported. Large supplies of what is known as American white ash are received in Great Britain from the States and used by consumers to supplement a deficiency in the supply of the native-grown wood. The timber is similar in quality, but has not the toughness or the all-round good qualities of English wood. It is imported in straight-grown round butts of moderate girth, and also in converted planks and boards, and is largely used for motor-body and wheelwrights' work, requisites for sports, such as hockey-sticks, tennis rackets and other articles. The tree, when cut down, has the habit of re-growth from the stool, and this second-growth wood is esteemed by some consumers as superior to the first.

From the Dominion of Canada a species of rather different character and known there as Black Ash, and in the United Kingdom as Quebec White Ash, was formerly largely shipped in squared logs, ranging from about 12 in. to 24 in. in diameter. There was a good demand at the time, as the wood was fashionable as a material for the cabinet-maker, being principally used in the manufacture of bedroom furniture. It was generally of clean growth, white in colour, and had, in good qualities when fresh felled, a silky grain which was greatly appreciated when used for the above purpose. The wood, however, did not retain its white appearance for any length of time after use, becoming brown with age, and gradually the fashion, after being dominant for a good few years, died away. The wood is lighter in weight than most species of ash, and is also less strong and tough. It possesses excellent working qualities, but except for the above purpose has been little used. Shipments at the present time are practically stopped.

Hickory.—There are said to be about a dozen species of this wood, mostly growing in a range of territory extending from the St. Lawrence to Mexico. The trees produce a generally strong, tough and elastic wood, which is largely used in the States. One species only, the Shag Bark Hickory—perhaps the most valuable— is shipped to Great Britain. The wood is of coarse, open grain, brown in colour, strong, fairly heavy and very elastic. It is used for the same purposes as ash, and is the wood generally employed for the handles of golf-sticks and, occasionally, for fishing-rods and similar purposes. It is generally exported in the form of round logs, but sometimes in converted planks, and the second growth, as with the ash, is more appreciated by some consumers than the first.

Walnut, Black or American.—The value of this fine wood was recognised as early as the seventeenth century, when large tracts of land were afforested with the trees. It has since then been recklessly exploited, and total extinction would be now well in sight but for slight schemes of re-planting which are taking place. The tree is of handsome growth, and attains to a height of 100 to 150 ft., with a diameter in well-grown specimens of 4 ft. and over. The wood produced is of more or less dark brown colour, even and uniform in texture, and has exceedingly good working and polishing qualities. Formerly, the wood was shipped in moderately squared logs, but during the last thirty years a continuous development in the export of square-edged, converted planks and boards has taken place. It was at one time extremely favoured for the making of bedroom suites, dining-room and library furniture, the fashion having been retained for many years, but, with a marked deterioration in quality and continual rise in value, its place was taken for these purposes by mahogany, oak and other woods. Besides its use in the cabinet trade it was also employed by builders for shop and office fittings, and for many other purposes.

The tree was also fairly abundant in certain parts of Canada, and regular supplies of an even better wood than the American at one time reached Great Britain, under the name of Quebec Walnut. It was superior in texture and colour, having a somewhat darker purplish tinge as compared with the American. It commanded a higher price, but the supply was apparently not large, and very little has been seen during the last ten years or more.

Whitewood.—Generally known in the States as Poplar, and in the United Kingdom as Whitewood or Canary Whitewood, the timber is really the produce of a Tulip tree, one of the most handsome and valuable that grow in the United States. It is, or rather was—for the supply shows unmistakable signs of exhaustion—plentiful in most of the Eastern States, growing to a height of 150 to 190 ft., with a clear bole, in exceptional specimens, of 80 ft., free of branches.

The wood was first shipped to British markets about thirty-five years ago in the shape of squared logs of extremely large dimensions, and quickly found appreciation, the trade developing, as in the case of most American woods, into the export of the material in a converted form, much of the supply being, moreover, dressed or planed on both sides. The wood is in most cases, and when of the best quality, of clear bright yellow colour, but varies, being sometimes of a brown shade and often of a greyish white. It is a clean, straight-grained wood, moderately hard, and works, stains, and polishes exceedingly well. It is not, however, durable, decaying rapidly when exposed or in a damp position, but is an admirable wood for indoor use, and popular with cabinet and pianoforte-makers, joiners and others who desire a cheap, easily worked wood that does not warp or twist, and that can be obtained in good dimensions.

Satin Walnut.—Known in the States as Red Gum, Sweet Gum, and Busted, but marketed in Britain under the name of Satin Walnut. This tree has its habitat in a somewhat similar area of country to that of the Tulip tree, but extends further into Texas, along the Gulf of Orleans. It is also, like that tree, of somewhat similar ornamental appearance and of equal size.

The wood is of fine, close and even texture,

greyish-brown in colour with darker stripy streaks. It warps and twists badly, and is most susceptible to decay when in damp places. It has, however, been tried for street-paving purposes, but with poor results, and is used for the cheapest class of furniture making, but is not popular, the low price at which it is placed on the market being its only attraction. The timber has much sapwood; this also is marketed, being known under the name of Hazel Pine.

Bass Wood.—This is identical with the Lime or Linden tree of England. Small parcels of this American and Canadian-grown timber, in converted boards and planks, are occasionally shipped to Great Britain. The consumption, however, is trivial, the principal users being the pianoforte action-makers.

Tupelo.—This timber is procured from a tree which passes in America under the name of Cotton Gum or Tupelo Gum, and is grown in swampy and occasionally water-logged districts in the Southern States. The wood is of greyish-white colour, close grained, clean and somewhat heavier than Whitewood. It is fairly workable, but is not durable in damp positions, and is inclined to twist and warp. A few shipments of converted boards are occasionally imported to the United Kingdom, but, although a low-priced wood, the consumption is extremely limited.

Maple.—This is known in the States and Canada as Sugar Maple, Hard or Rock Maple, and also when figured as Bird's-eye Maple. The tree is largely grown in Canada and also, to a much less extent, in the United States, and produces an important and useful timber. It attains a height of about 100 ft., its trunk developing a diameter of 3 to 5 ft. The wood is of yellowish or greyish-brown colour, is extremely hard and close grained in texture, and heavy in weight. It works well and finishes

with a satiny surface. A moderate quantity is imported
into Great Britain principally in converted planks and
in the form of floor-boards, ready dressed and prepared
for laying down. The chief use of the wood is for this
flooring purpose, and most of the skating-rinks which
sprang into existence a year or two ago were covered
with this fine hard wood. A fair amount is also used for
turnery purposes, and some little of the wood in the
making of printers' joinery. The timber is sometimes
of figured character, so well known as Bird's-eye Maple.
This is supposed to be some freak in the development of
individual trees, probably arising, it is surmised, from
the suppression of dormant buds. Large quantities of
the wood were formerly used in the shape of veneer,
for picture-mouldings and other decorative work, but
beyond a few logs occasionally shipped comparatively
little is used at the present time.

Cotton Wood, or Carolina Poplar.—This tree, the
largest representative of the genus, is more or less well
distributed over the Eastern States of America.   It
is stately in appearance, surpassing in many cases 100 ft.
in height, and producing a light, soft, willow-like wood
that can be extracted in large dimensions. Occasional
parcels of the wood have been tried in the English
markets, but no great use has been found for it.   On
the Continent, however, principally in Germany, it
is in some favour.

Persimmon.—The Ebony of America.   In its growth,
this tree attains a height of 40 or 50 ft., but specimens in
favoured localities in Indiana and Illinois reach greater
heights, and grow to a diameter of from 2 to 3 ft.   The
wood is heavy, dense, and close-grained, the heart-wood
of old specimens showing a dark, almost black colour.
Most of the wood exported, however, is sap-wood of a
greyish-brown colour.   The limited amount that is

exported, principally received in the Liverpool market, is used for shuttle-making, turnery, and other purposes.

Rock Elm.—A valuable tree, which in some respects is akin to Scottish and English Wych Elm, is principally found in Canada and the Northern parts of the United States. Growing to a height of 80 to 100 ft., with a trunk of 3 to 4 ft. in diameter, it inhabits the upland portions·of the country, preferring rocky ridges and slopes. The wood is heavy, very fine and close in texture, and of a light-brown colour. It works well, and is greatly appreciated for its strength, toughness and flexibility. Like most Elms, it is not liable to split, and is exceedingly durable when in wet situations. Principally exported in squared logs, it is largely used by boat and ship builders, and also for the construction of bent-work for wheelwrights.

Orham.—This is an unidentified species of Elm which is exported in small quantities, and principally used for coffin construction and wheelwright purposes.

Birch.—Seventeen varieties of this wood are said to have their habitation in Canada and the United States. Practically only one, however, is exported, being principally shipped from Quebec and other Canadian ports. The tree from which this timber is produced reaches a height of 70 or 80 ft., and has a trunk diameter of from 2 to 5 ft. ; the best specimens are obtained from well-drained, upland localities. The wood is of light brown or light reddish-brown colour ; is hard, fine and close in texture, heavy and strong. It is not durable in damp situations, and, although working well on the whole, is occasionally very cross-grained. A large import has taken place for many years to Great Britain, principally in the form of squared logs, and also of late years in the shape of converted planks

and boards. It has many uses ; formerly it was fashionable for the manufacture of the cheaper classes of
furniture, but the consumption at the present time is
principally for turnery purposes, chair-frame making,
coach-building, and many further minor uses.

CHAPTER XIII

A RICH wealth of valuable woods is to be found in the above regions. Large areas, notably Brazil and some of the Northern parts of South America, are practically unexploited in this respect, and, although a fair number of specimens are well known, many more still remain that are awaiting recognition. Over 400 species in Brazil alone are said to be known at the present time, most of them, however, being unidentified. Previous mention has been made in a separate chapter of the mahoganies with which the districts of Central America are so richly afforested, and it now only remains to make some few observations on the further species of timber which are obtained from these tropical districts, and also from the further areas of South America.

Satin Woods.—From most of the West Indian Islands this beautiful wood is obtained in more or less quantities and in varying qualities; the finest, however, is shipped from St. Domingo and Puerto Rico. The wood is of a clear, bright yellow colour, dense, hard and heavy, and at times is beautifully figured. It has for many years past been held in high estimation as a material for cabinet-making, and some of the much sought-after pieces of furniture of the latter part of the eighteenth century were partly, or wholly, veneered with wood which doubtless had its origin in the Island of St. Domingo. A fair supply is still shipped from this Island in squared logs up to about 15 or 16 in. square,

114

but figured specimens are scarce among the number. Such logs always command high prices, especially when of finely figured quality, and there is a record of one making 21s. per foot in public sale in London some few years ago. The figured logs are still used by the cabinet trades, but to less extent than formerly, the bulk of the total imported into the United Kingdom at the present day passing into the hands of brush-makers, for the making of the backs of expensive hair-brushes. The wood found in the adjacent Island of Puerto Rico is held perhaps in greater estimation than the St. Domingo wood, being of a brighter and richer colour. Unfortunately, however, this supply has become practically exhausted, and very little besides roots and rough stumps of trees is shipped at the present time. Even these are in demand, notwithstanding the large amount of waste that occurs in their conversion. Pieces have been known to sell at £110 per ton, the usual price, however, being about an average of £10 to £30 per ton.

Lignum Vitae.—This is a wood which is common to several of the West Indian Islands, the tree from which it is obtained growing perhaps more freely in St. Domingo and Cuba than in the other islands. It is the heaviest wood in existence, its dry weight being approximately 80 lbs. to the cubic foot ; even a piece of its thin covering of bark will sink when immersed in water. Its great weight indicates its extremely dense character, it being perhaps one of the hardest woods existing. In colour it is of a dark brown, streaked with black, the sap-wood, of equal value to the heart-wood, being yellow and of very narrow proportions. The wood is of close texture, and is largely impregnated with a sticky resinous or gummy substance. It is exported in short lengths and of generally small dimensions, rarely exceeding 10 in. ; occasionally larger pieces are shipped, but they are almost invariably subject to cup shakes which spoil their value. The wood is principally used for ship-blocks, also to some extent for rollers, for the making of bowls, and for turnery and other miscellaneous purposes.

Quassia.—This wood is the product of a tree which has a fairly wide distribution in the West Indian Islands, Brazil, and other parts of South America. It is a rather soft and stringy-grained wood of light yellowish colour, and, beyond the extremely bitter taste which it possesses, has no other feature to distinguish it. Small parcels of rough unmanufactured pieces of this wood are, from time to time exported, the consumption being mostly for medicinal purposes and for the making of extracts for spraying hops and other horticultural produce.

Lance Wood.—From Cuba, Jamaica, and other Central American districts, but principally from the first mentioned Island, supplies of this wood are obtained. It is a heavy, dense, close-textured wood of yellowish colour, its chief characteristic being its extreme elasticity and toughness. It is rarely seen of large size, and, shipped in what are known as spars—lengths of about 12 to 16 ft., and about 6 in. to 8 in. in diameter—finds its chief users among benders of timber for shafts of vehicles and other purposes in coach and wheelwright work. It was formerly rather extensively used for fishing-rod making, but has now been altogether superseded by green-heart for that purpose.

Degame.—A quite distinct variety of the same wood as the above. It is similar in most respects to the foregoing, shipped from the same sources, exported in the same fashion, but in rather larger spars, and is used for identical purposes.

Blue Mahoe.—This timber has a fairly wide distribution in the West Indian Islands and in adjacent countries. The wood is of greyish-brown colour and, otherwise than for colour, might be compared in weight, strength and flexibility with English Ash. A few logs in squared form are occasionally met with, principally exported from Cuba, but the consumption is quite negligible.

Washiba.—This is another timber of the same tough and elastic character as the foregoing, and is obtained from Guiana and other sources in South America. It is reported as obtainable in long lengths and of large dimensions, but is somewhat scarce. No great amount has been shipped at any time. The wood is of reddish-brown colour, hard, heavy, strong, and very flexible.

Tulip Wood.—A beautiful fancy wood of small growth which is obtained from the forests of Brazil, is known under this name. The wood is fairly heavy, hard and close textured ; its striped appearance bears some resemblance to a pink and yellow-striped tulip, hence its name. It is not at all common, the small amount exported being principally used for the making of bandings in marquetry work and for fancy turnery purposes.

Partridge Wood.—This is another fancy wood obtained from the same source as the Tulip wood. It is extremely heavy, hard and dense, of a very dark-brown colour with darker markings, and is used for similar purposes as the above. In appearance it is somewhat similar to Cocus wood, of which an account follows, and is often mistaken for that wood.

Cocus Wood.—Also known as West Indian, Brown, Jamaica Ebony, and by other names. This wood is well distributed in America and the West Indies. It is an even heavier wood than the foregoing, ranging from 77 to 87 lbs. per cubic foot ; compares with Ebony in hardness, and is of a rich, dark-brown colour.

It is exported in small pieces of about 6 ft. to 8 ft. long, and from 3 in. to 6 in. in diameter, which have exterior sap-wood of about ¾ of an inch in thickness, and is used for the manufacture of flutes and other musical instruments, small cabinet work, and turnery purposes.

Purple-heart.—This is a Demerara fancy wood occasionally met with, and is obtained from a tree which grows to large dimensions, both in height and girth. The wood is hard, heavy and strong, and its appreciation is due to its colour, which, brown when first cut, afterwards turns to a decided purple. This rarity of colour renders it an acquisition to marquetry cutters and to the cabinet-makers.

Sabicu.—A Cuban wood which is in favour for its strength and durability among naval shipbuilders, ranking high in Lloyd's list. It was formerly largely used in Government work, but is now little required, the use of iron, and alterations in methods of construction, having rendered its use a thing of the past. The wood is of rather dark-brown colour and is fairly hard and heavy, and, exported in squared logs of similar size and make to mahogany, is sometimes used in place of that wood. It is also often finely figured like mahogany, having, but for its brown colour, a very similar appearance.

Crabwood.—Another hardwood which is principally supplied from British Guiana, but which is obtainable in other adjacent countries. This is a moderately hard wood, somewhat coarse in the grain, and of reddish-brown colour. It is extracted in large squares and of long lengths, but the export of late years has been principally in the form of converted planks and boards. It is shipped as a probable substitute for the cheaper kinds of mahogany, but, although selling fairly well when the regular descriptions are scarce or their price high, is not appreciated to any great extent.

Green-heart.—This wood is native to British Guiana and species are found in adjacent parts of South America and in the West Indian Islands. It is an invaluable wood for many purposes, and ranks as first-class in Lloyd's list of shipbuilding timbers. It has a recorded dry weight of 64 to 75 lbs. to the cubic foot, and is of a brown colour, with an occasional greyish-green cast. It is extremely strong—one of the strongest in existence—and is unequalled in its flexibility. The chief virtue, however, which it possesses is its resistance to the teredo worm when used for piles and other work in sea-water, being unsurpassed in this particular by any other wood. It is exported principally in the form of moderately squared logs, which range up to extreme lengths, and is largely used for the above-mentioned sea-coast work, and also in the building of bridges and for other constructional purposes where durability and strength are needful. A minor use is the manufacture of fishing-rods, the flexibility of the wood making it an ideal material for the making of fly-rods.

Mora.—Another fine wood from British Guiana, which is produced from one of the most valuable and also one of the largest trees which exist in that possession. It has been at times largely used in ship construction, being rated like Green-heart as first-class. Unlike that wood, however, it will not resist the teredo worm. It is obtainable in large dimensions and the wood, dark, reddish-brown in colour, is heavy, hard, strong and most durable. It is said to have poor working qualities, and is only fit for heavy constructional purposes. Very little has been used since iron took its permanent place for the building of ships.

Snakewood.—A choice tropical wood principally obtained from British Guiana. This is one of the most appreciated woods for stick-making that are known. The wood is of fine, close grain, and of rich brown colour, mottled in a curious way with black rings. Extremely high prices have been paid for this wood at times, and owing to its scarcity it commands high rates at the present, when, perhaps, not so much in fashion.

Rosewoods.—Besides the East Indian variety of this wood, which is mentioned in a former chapter, two further species are extracted from Brazil and Honduras respectively, and shipped to European markets. Others, besides, are found in various districts of Central America, but are only rarely seen. The Brazilian variety is the finest and the one which was chiefly employed when this wood was fashionable for the manufacture of furniture some forty or fifty years ago. It is usually shipped in rough pieces, more often in the form of a trunk, roughly cleared of sap and split down the centre. The wood is very hard, heavy, close-grained, and of a rich and very dark purplish brown ; the best is that shipped from Rio de Janeiro, being finely marked with black or very dark-brown stripes. Further wood of a similarly good character is also exported from Bahia. The Honduras wood is principally exported from British Honduras in short, round billets of about 10 in. to 15 in. in diameter. This wood is almost as dense as the Brazilian variety, and is of nut-brown colour, streaked with black stripes. Both woods are much less used than formerly, there being very little demand for them from the cabinet trades, as the wood for that purpose is quite out of fashion ; it is also little used for piano case-making at the present time, but some amount, in the shape of veneer, is still consumed.

Log Wood.—An important dye-wood which is extracted from most Central American districts, the principal source of the supplies however, are British Honduras and ports on the Eastern Coast of Mexico. It is shipped in irregular and ugly-shaped pieces of root and branches, from which are obtained red and black dyes. In alcohol it gives a beautiful greenish-yellow colour which turns to an intense crimson on the addition of potash. A solution in water gives a port-wine colour, and it is a matter for conjecture as to whether there is any connection between this and the " wine from the wood " which the wine merchant recommends; if so, the fact is never acknowledged. It is said to be used extensively in the making of ink as well as for many other dyeing purposes.

Fustic.—Another dye-wood which is exported from Central America and Brazil. This wood is of fairly hard description, having a fine, satiny texture, and being of yellow or yellowish-green colour. Occasionally, logs of small size are to be seen, but generally the wood is shipped in small pieces of irregular and rooty growth. The timber is used in Great Britain solely as a dye-wood.

CHAPTER XIV

SOME NOTES ON THE EXTRACTION OF TIMBER IN VARIOUS COUNTRIES

THERE are few perhaps that realise, on looking around on the different varieties of wood that surround us, the amount of labour that has been expended on the production or the vicissitudes through which these timbers have passed in the long journey from their native forests to their present location. They probably have no conception of the far away forests, sometimes hundreds of miles from the confines of civilisation, from whence the timber is obtained, nor of the arduous tasks of felling and transporting it by various means to shipping points ; they know little about how the latter work is effected, sometimes under conditions of almost Arctic severity, and at other times in tropical latitudes, where from dense-tangled jungles, practically impenetrable even to sunlight, the stifling and moisture-laden heat is almost unbearable to any but native labour.

From the simple and primitive means of extracting the trifling amount of timber which is obtained from English woodlands to the methods in vogue in America, Canada and other countries, there is an extensive range, and a brief account of one or two may be of interest.

In most countries the natural and first-used means of transporting the timber from the forests to more convenient places for conversion and shipment is by the utilization of the waterways. Areas of forest land contiguous to the main streams are first exploited, and, as these become exhausted, further and further afield do the timbermen have to travel, until tributary

122

streams, sub-tributary and waterways of all kinds are brought into use, water having to be impounded in many of the latter to provide sufficient current when needful at certain seasons.

In the vast Dominions of Canada, which have yet to be covered with a network of railways, this natural and economical method of bringing the timber nearer to the coast is universal, and the subject of the arduous tasks of the lumbermen in felling the timber in the woods in the depths of winter, the manipulation of the drives in the snow-flooded streams in the Spring, and the life led by these lumbermen, have formed material for much picturesque writing by various authors.

The camps are formed in the late autumn months, some fifty or sixty men, mostly French Canadians, being included according to the area of the " stand " which is to be operated upon. Shacks are erected, and stores of all kinds needful for the men's subsistence during the ensuing months are provided. Through the depths of the winter, far away at most times from civilisation, with snow lying around to the depth of six or more feet, and with the thermometer registering ten, twenty, or even thirty degrees below zero, the work of felling the trees in the silent woods is carried on. The trunks, after being felled and branded with a hammer-mark to identify their ownership when they arrive later at the mills, are then skidded down the hill-sides or drawn by horses over the frozen snow to the banks of the waterways, where they are piled in dumps in readiness for the awakening of Spring.

After completing their arduous spell of winter's work the men, who are not notorious for their peaceable habits or the steadiness of their lives, are paid off and return to civilization in the populated centres of Ottawa,

Montreal and other districts.   Like the common habit—
or formerly common habit it may be written—of seamen
to dissipate their well-earned wages on returning from
a voyage, these lumbermen by reckless dissipation,
or what is popularly known among them as " Blowing
it on," quickly rid themselves of their plentiful stores of
hard-earned money.

In the early months of the year, or at a period when
the grip of winter shows signs of relaxing, another
gang, whose business is to see to the floating and driving
of the logs, is formed and proceeds to the woods.   Of
all the operations connected with lumbering, from the
first cutting of the timber to its arrival at the mills,
no part is so full of anxiety to the extractors as this.
The flooded streams have to be utilized while they are
in this condition, or there is a certainty of logs being
hung up, with a consequent loss of timber, loss in
expenditure on labour, and a shortage of supplies to
keep the mills at constant work.

The huge piles, or dumps, on the river banks are broken
up, a tool called a pelvie being largely used for this purpose.
The Canadian dollar bill has on its face an illustration
of lumbermen at work, the men having this tool in use.
Frequently, at the commencement of the driving season,
the huge dumps are so firmly frozen together that
human efforts are unavailing, and recourse is made
to the use of small charges of dynamite to break them
up.   This done, they are tumbled into the ice-filled
streams and started on their voyage towards the sea.

As previously mentioned, some of these tributary
waterways or creeks that are connected with main
rivers or lakes do not hold sufficient water to secure a
passage for the logs, even in flood time, and dams are
constructed to hold up the water.   At these spots
gangs of men are usually located, guiding and steering

the logs as they collect down the sluices and slides into the rapids below, so that no obstacle shall cause a jam. The work, especially in breaking up any jams that occasionally occur, is most arduous and risky, carried out as it is in strong-flowing streams at an icy temperature. To see these men at work, jumping on and manipulating ice-covered logs in turbulent rapids and strong streams, is a sight worth going a long way to see. It is an exhibition which shows a high development of nerve, skill, agility, and daring ; the risks are great, however, and many accidents occur that occasionally lead to fatal results.

The winter-piled dumps are gradually cleared, and a general sweep up of the watercourse takes place after the great mass of the logs has gone down. Stray logs, hung up in bends and creeks, are diverted into their proper onward course, and by the commencement of summer the work is so far completed.

The labour of felling the lumber in the winter, and the further handling of the logs in the icy Spring waters, is trying enough, but is preferable, if we can believe the stories of many who have had experience, to the work which is completed in the early months of the Summer, when the thermometer rises the other way and up to 150 degrees above zero. The fly season then commences. According to the above authorities no mortal can adequately picture the systematic and methodical way in which these insects torment and make miserable the life of man. There are black-flies, sand-flies, deer-flies, and mosquitoes, and they work, day and night, with monotonous perseverance in their onslaughts. They appear to have some system in their assaults, the different species working in regular day and night-shifts, not in forcible frontal attacks, but in stealthy guerilla assaults which leave their victims covered with a mass of clotted blood marks. The mosquito is described as rather harmless and gentle in disposition in comparison with the sand-fly, whose bite is represented as similar to the thrust of a red-hot needle—clothes, blankets and rugs being non-impervious to this insect's attack.

From the tributary streams the winter-felled logs find their way on the currents to main rivers, where, under the guidance of the drivers, they are taken down towards the mouths where mills are situated. Here the driving gang are paid off, and the men, who may be described as picturesque in the abstract, continue to " blow their wages " in a similar manner to those first employed.

Many of the most important mills are established at Ottawa, this town being undoubtedly the metropolis of the Canadian timber trade. From the great watershed of the Ottawa river vast quantities of pine and spruce are annually brought down to these mills, where they are sorted for the different owners, converted into planks, deals, boards and other sizes by modern machinery, principally worked by the power of the fine falls which exist on the river, and are then transferred on the St. Lawrence to Quebec for shipment to the different markets of the world. As showing the extent of the trade, it may be mentioned that the output from the mills at Ottawa and the neighbouring town of Hull on the other side of the river, is estimated at about 300,000,000 ft. of timber per annum.

The network of railways that intersect the United States, after a longer period of development, makes the method of extracting timber in that country somewhat different from that which obtains in the neighbouring Dominion of Canada. The active hustling enterprise of the Americans is seen on all sides, even in this industry.

Photo by
Darius Kinsey, Seattle, Wash.

TIMBER-HOISTING

Up-to-date machinery is taken to the woods, if the stands are sufficiently extensive, and logging is here carried on under altogether advanced conditions. Roadways are made, metals are laid down for trucks and locomotives, aerial wire transporters are largely used and, generally, the logs are converted into lumber, dressed, dried and prepared for market on the spot, the goods being transported to the nearest railway for conveyance to home markets or to the coast for shipment to other countries.

In some districts of the country, however, where railway systems have not so fully developed, other methods of extraction are employed, and an interesting one that works a district on the Western slopes of the Rocky Mountains in California is worth noting.

The timber from a large district on the steep sides of these mountains is brought down to mills near to points of exportation by means of what is known as a shute. The work of constructing this shute, or flume as it is sometimes called, is said to be a fine example of engineering skill, traversing, as it does, hill-sides, ravines, valleys, streams and other obstacles in its course, which, including side-branches, is 125 miles in length, and is still being continually extended. It is built of wood in the form of a trough which, starting at the top of the mountain, is carried in various directions convenient to the timber which has to be transported. The trough, or gutter, is laid at a varied gradient, and the logs are placed inside, water being turned in from a reservoir at the head of the shute. On this current the timber is transferred at a velocity which varies with the gradient, at some places rushing along at 20 miles per hour. At various stations, chiefly where side-tracks join the main flume, men are stationed to prevent jams and to keep the course free of obstacles. To show the extent of this undertaking, and the amount of timber that is brought down, it may be stated that the actual cost of erecting this shute was over £1,000 per mile, and the company who work the concern handle an output of something like 575 million cubic feet of wood annually.

It may be further noted that the original idea of the water-shutes that were at one time popular in America as a pastime and, subsequently, in other countries, was taken from this method of transporting timber.

After the few remarks as to how the Canadian forests are exploited, and the methods in force for transporting the timbers to the mills, there is little to add in regard to the system by which the vast quantities of coniferous timbers that are obtained from the forests of Russia and other Northern European countries are prepared and shipped to markets. The methods are practically identical; both railways and roadways are inadequate, especially in Russia, and, as in Canada, the waterways are largely used. On the currents of the rivers the logs are transported to the mills, which are placed at ready points in the Baltic, Gulf of Bothnia and the White Sea. These mills are equipped with up-to-date machinery suitable for converting the timber into dimensions required by different markets. Most of the ports in these regions are, however, ice-bound during certain months of the year and, consequently, exports are only made during the open-water season.

Various other interesting means of producing timber from its native source, in relatively small quantities for native consumption, obtain in various parts of Europe. In some districts of Austria and Germany, where the configuration of the land admits, slides with more or less steep gradients are common. These slides, or sledges, are either drawn by horse-power or worked by men, and are constructed of two horizontal pieces of wood turned up at the front and shod with iron throughout their length, cross-pieces connecting the two shafts. On this frame the logs are placed crossways, and with an efficient brake to control the speed the man in front can bring down, it is said, a load of 55 to 70 cubic feet, the gradients of the track being from 1 in 14 to 1 in 4. Short lengths of timber, suitable for pit-props and other work, are generally transported by this manual labour, but longer lengths are moved by horse-power on more gradual slopes.

A high development in the use of the waterways has taken place in these latitudes, principally extending from the Black Forest eastward to the Bavarian Alps in Austria. Extensive schemes for rendering the small tributaries adaptable for floating the logs have been undertaken, so that it may be possible for the timber to pass down these mountain streams to the broader rivers. Here they are formed into huge rafts, some of those to be seen on the Rhine being frequently 600 ft. in length, and with a breadth of 100 to 150 ft. Steered with long sweeps, similar to those in use on the Thames lighters, or by rudders when they are towed on slow-moving currents, these rafts convey the timber in an economical way to many markets along the course of the rivers. Huts erected on the logs accommodate the small crew, the raft being also used for the transport of bulky and raw produce of all descriptions. In the Hungarian portion of the Austrian Kingdom very modern means of exploiting the forests are used, most of the coniferous wood shipped from Galatz having been transported from the mountainous forest slopes before conversion, by cable lines, funicular railways and other up-to-date means of transportation.

The subject of the extraction of hardwoods from tropical countries is perhaps as interesting as that relating to the production of coniferous woods from more temperate latitudes. In these countries approaching the Equator, Mahogany and Teak are the principal woods obtained, and of the latter mentioned some observations have been made in the chapter devoted to Asiatic woods.

Most of the countries from which Mahogany is obtained are in no very advanced stage of development, and there is, therefore, a lack of means of transport, Consequently, in the absence of roads and railways, waterways have to be utilized, rivers, full with seasonal rains, and much native labour, being necessary to exploit the forests.

In Central American districts where Mahogany is found, somewhat primitive means are employed for its extraction. A few, however, are being exploited under the restless energy of American firms, but, generally, they are worked under similar conditions to those that have been in use from the first. Practically all the labour is done by natives, chiefly by negroes, who, in addition to the task of getting the timber from forest to stream, are also employed in squaring the logs for market. In many districts of Central America, but especially in British Honduras and Mexico, they have attained a high degree of efficiency in this latter work, straightening and squaring huge tree-trunks into most shapely and well-made squares, the work being simply performed with axe and adze. The floating of the logs to the mouths of the rivers during the wet season in June or July depends upon the amount of rain that falls, and, should there be a deficiency, many logs are sometimes held up in the by-ways of the river, there to await better conditions during the following season's rains. As in these moist, humid, tropical countries, decomposition is rapid, the effect is sometimes shown on the wood, at times in regard to its appearance, at others, by the attacks of small worms ; with a consequent depreciation in its value.

Interesting as this work is in the Central American districts, it is perhaps more so in regard to the extraction of this wood from the dark and comparatively little known forests of the West Coast of Africa. There is a mysterious glamour about these forest regions which attracts interest, for little is known about them owing to their vast extent, the wonderful wealth of woods they contain, and the short period of time since a dark space marked their locality in our geography books.

Those who come in contact with the wood from these regions, whether in the log as it has been shipped ; in a converted form as it reaches the consumer ; or as furniture in the office or household, have little conception of the amount of labour entailed nor the many vicissitudes through which the wood has passed since it grew in lordly pride as a monarch perhaps among other huge trees in some dense, dark, and little trodden forest. Very often the tree from which a log has been produced has grown hundreds of miles from the coast, perhaps two or three miles from a waterway, and as the transport has to be performed entirely by human labour, and the logs weigh up to about 5 tons, it may be imagined the amount of toil and endurance which has to be expended in this steaming tropical climate before the log reaches the first stage of its journey to the various markets of the world.

Europeans of English and French nationality, and also one American firm, are principally interested in the outget of this wood, and have their agencies at many places which serve as ports along this great extent of territory. All the labour in the forests, however, is done by native " boys " with, in many cases, white superintendents over them.

In the British possessions in Southern Nigeria, most of
the forests are fortunately under the jurisdiction of a
Government Forest Department, and on a concession
being obtained for cutting the wood in a certain area
the lessee binds himself to certain conditions—the size
of the trees to be felled, and the replanting of others
to take their place being important provisions.   This
licence being granted, the trees have to be located,
and as Mahogany does not grow in stances or clumps,
but scattered to an extent of sometimes a quarter of a
mile apart, this difficulty in a dense, almost impenetrable jungle, is met with at the onset; moreover, in accor-
dance with the Government regulations, no trees must
be felled that do not reach a circumference of 12 ft.
at 10 ft. from the ground, any offence against this
condition being punishable with a fine.   The trees
located on the area are marked, and after a government
supervisor has inspected and a felling charge has been
paid—a part of which goes to the paramount chief of
the locality—a further permit to fell these special trees
is granted.   These mighty trees, often towering to a
height of 200 or more feet, and occasionally with a girth
of 36 ft., have usually at the base, to support their
growth, natural buttress-like projections which extend
from 6 to 12 ft. upwards.   A rough stage is erected
by the natives over this supporting growth, and then
commences the labour of felling the tree.   It is entirely
done by axe work, as many " boys " as can be got to
work on both sides of the trunk being employed.   If
the tree is an upright one in its growth, the chopping
continues until only a few inches of the wood is left in
the centre of the tree, when one last blow or a gust of
wind upsets its equilibrium, and with a crash that
makes the earth tremble for a distance of 200 or 300 feet
and which brings down other trees of less substantial

growth, the monarch of the woods is laid low. The trunk is then measured into suitable lengths, mostly from 12 to 24 ft., and in most instances logs are also obtained from the branches of the tree, it not being an unusual occurrence to get ten out of one tree, although four is about the average number. The sections of the tree are then squared, with one end tapered or shaped to facilitate hauling; and after numbers and marks have been painted on each log, and a Government pass-mark likewise imposed, the logs are ready for transport to the rivers. Roads are cut through the bush to the nearest creeks or waterways, and on this path small round trunks of trees to act as rollers are placed. A wire rope is passed round the log, and to this is attached a further length having, at intervals of about 3 ft. apart, pegs of wood inserted. With a hold on these the gang of perhaps 80 or 100 men commence with a sing-song chant their task of hauling these logs, with an average weight of perhaps 3 tons, to the water-courses. Here, if they are in creeks where there is little water, they have to await the rainy season, and on the flooded currents they find their way into the larger streams and, ultimately, to the coast.

Very few of the so-called ports along the West Coast have good loading facilities, owing to the nature of the seaboard, and the logs are therefore, in many cases, towed out to the boats that lie off the coast, being lifted by the special and powerful appliances these vessels possess, straight out of the water into the holds. A further stage in the history of the logs follows. They are transported by these vessels to a port—probably Liverpool—where they are first yarded, then included in a catalogue with many others, and afterwards sold by public auction. They may possibly be purchased for re-transportation over the Atlantic to American

By permission of the   " Timber Trades Journal."

INITIAL OPERATIONS ON A BUTTRESSED MAHOGANY TREE

consumers, or they may be bought for home requirements, the merchant who has purchased re-selling them to yard-keepers, who convert them, and, after keeping them many months for seasoning purposes, eventually sell them to manufacturers, by whom they are used in making the many articles that are to be seen around us.

CHAPTER XV

WITH the advent of iron, particularly in its adoption for shipbuilding, in its use for girders and framings for buildings, and for innumerable other purposes, it was long ago foretold that timber would not be necessary to the same extent in future, and the foreboding was again freely expressed, but in a somewhat less positive manner, on the introduction of re-inforced concrete as a constructive material. Notwithstanding all these prophecies, however, timber has emerged triumphantly; it has withstood all competition, and, although the demand has fallen away, or lapsed entirely in some directions, other channels for its uses have been discovered, and the volume of consumption has grown year by year in magnitude, in all countries where civilization has progressed.

As showing this growth in the demand it is estimated, on good authority, that, in England and the United States alone, the consumption has doubled during the last half century.

The question therefore arises how this ever-increasing demand can be supplied. A timber famine has been talked about in a casual way for years past, but little heed is paid to the matter, the subject being a sort of perennial one that springs up from time to time when a shortage occurs. Varieties at intervals fail to meet the demand, and gradually cease to arrive on the markets, but others are found to take their place, and in time become established; and although there are fluctuations in the imports, due to labour disturbances, weather conditions, and other extraneous circumstances, no pronounced lack of timber has as yet occurred.

The pressure of this demand is, however, increasing and becoming more widespread year by year. Exporting countries have relapsed, or are on their way to become importing countries, and there is, consequently, a continuously increasing drain, with no corresponding augmentation in the estimated supplies.

The position in regard to Europe, as shown from statistics that cover an average of five years, and which are given on the authority of Sir William Schlich, is that only five countries on the Continent exported timber namely—

		Tons
Roumania .	. . .	60,000
Norway . .	. . .	1,040,000
Austria-Hungary	. . .	3,670
Sweden . .	. . .	4,460
Russia and Finland .	. .	5,900,000

While thirteen others imported—

Great Britain .	. . .	9,290,000
Germany	. . . .	4,600,000
France .	. . . .	1,230,000
Belgium	. . . .	1,020,000
Denmark	. . . .	470,000
Italy .	. . . .	420,000
Spain .	. . . .	210,000
Holland .	. . . .	180,000
Switzerland	. . . .	170,000
Portugal	. . . .	60,000
Bulgaria	. . . .	50,000
Greece .	. . . .	35,000
Servia .	. . . .	15,000

These figures giving a net import into Europe of 2,620,000 tons.

Other countries in all parts of the world are also more or less importers, a few having a total absence of native supplies ; others, while exporting some varieties, import others in greater quantities—principally timbers of coniferous growth for building purposes ; among those who entirely import are—

		Tons
Egypt		200,000
Cape of Good Hope . . .		150,000
Natal		50,000
Mauritius		20,000
China		50,000

While among those that export some varieties, but import others in larger quantities, are—

		Tons
South America		330,000
Australia		160,000
Ceylon		10,000
Japan		5,000

The following countries export entirely, or their exports are in large excess of their imports—

	Tons
West Indies, Mexico, Honduras, etc.	13,000
West Coast of Africa. . .	28,000
India	55,000
United States	1,020,000
Canada and Newfoundland .	2,144,000

It may be gathered from these few statistics that there is a constant drain upon timber resources in progress in all parts of the world.

Areas of forest lands in close proximity to ports of shipment have long since been exhausted, and further and still further afield the exploiter has to go to obtain his supplies, the cost of extracting these growing with the extended radius. Values have consequently been on the up-grade for many years past, data showing that in the United Kingdom, in the period between 1550-1750, the cost of timber quadrupled, and has since shown a continuous and progressive advance. In the whole of Europe statistics reveal that between 1894-1902 values appreciated no less than 20 per cent., and that in Russia, notwithstanding her vast native supplies, a proportionate rise took place. Sweden and Norway, also with supplies at their doors, experienced similar conditions, an appreciation of 15 to 20 per cent. having taken place during thirty-five years in the value of their native woods.

The effects of the great onslaughts that have been made on the virgin forests of the world, without any serious attempt at re-afforestation, are gradually being felt, and, as the conditions of an increased demand and a perceptibly diminishing supply continue, the outlook is certainly not one that can be regarded as satisfactory.

Another result of the clearance of forest areas which is in progress all over the world is the marked influence effected on natural conditions. As is well known, the woods mitigate extremes of temperature ; have a marked effect in regulating the water supply—more especially by insuring the sustained feeding of springs and thus rendering the flow of waters in rivers more continuous, and in tending to reduce the danger of violent floods ; they assist, by the action of the roots and stems of the trees, in preventing landslips, the erosion of hill-sides, and arrest the progress of shifting sands ; they, moreover, act as wind-breaks for the protection of agricultural areas and other necessary uses which are more perhaps to be noticed in tropical regions. Many countries that have been more or less cleared of woodlands have become sterile and arid wastes as a result of altered natural conditions.

Some of the West Indian Islands have slightly felt the result of denuding their forest lands, an instance of which may be mentioned. The State authorities that were in power some years ago in the Island of Jamaica were of the opinion that conservation of their abundant forests was unnecessary, as timber could be imported at very cheap rates from the United States. They found, however, after the woodlands of the Island had been practically exhausted, that some of the conditions above stated began to prevail. Their water supply began to be interfered with, there was no storage for the copious season's rain, rivers at times became torrents, with agricultural lands in the valleys submerged, while at others the river beds contained but a mere trickle of water.

Feeble attempts have been made in various countries to make some provision for the future in regard to timber supplies, but the efforts generally have been trivial and altogether incommensurate with the ravages that are being made on existing supplies.

Comparatively little has been done in the way of afforestation in Europe : Germany leads the way with a scientific and well-organised system that provides her consumers with a good proportion of their needs : France, too, has a systematic cultivation of forest lands, and the excellent results which she has attained in the planting of large areas of sand dunes and waste spaces in the Landes is an object lesson in economic providence. These formerly arid wastes are thickly planted with the Cluster or Maritime Pine, which, besides finding employment for many people, provide resin, from which most of the turpentine used in France is extracted ; the trees, moreover, when exhausted by tapping, are felled, and with the thinnings and loppings are shipped to South Wales, where they form a large proportion of the props used in the mines of that great coal centre. With the exception of these two above-mentioned countries, there is practically no other in Europe that conserves on any extensive or efficient scale.

In the East, the great Empire of India has, under the able management of an Indian Forestry Department, established a most efficient control over the present forest resources of that country, the systematic felling and control of the timber, and the organization of re-planting with a view to a continuity of future supplies being an example of State management which might well be followed in other countries.

Japan has an excellent and old-established system of State regulation of her forests, but few other countries in the East have taken the question in hand.

Slight attempts have been made from time to time in the various divisions of Australia to arrest the decline of woodlands, but on no scale of importance, and in New Zealand there has been little movement in that direction.

On the African Continent, France conserves much forest land in Algiers and Morocco, the area being reported as something like 5,000,000 acres, and controls, it is understood, the timber in her possessions on the Ivory Coast and in the Congo Territory.

On the same West Coast the Southern Nigerian possessions have been placed under the control of an experienced Forest Conservator, with the object of securing reports on the forest resources of the country ; and to prevent excessive exploitation with a view to continuous supplies. Doubtless the valuable mahogany and other timbers of this locality will now be secure from reckless cutting, and replanting will systematically replace the timber that is exported.

In the great districts of the southern parts of Africa— Cape Colony, the Transvaal, Natal and other divisions, there is a lack of timber, especially of that suitable for general building purposes. Very large quantities might be grown for the use of the community and for the benefit of agricultural areas, but little interest apparently is ever taken in such a question.

Turning to the Western hemisphere, attempts are being made in British Guiana to preserve some of the valuable woods of that country, and probably the timbers of the French possession in the same locality are efficiently looked after. Nothing, however, is heard of any attempt being made to control the outputs of other States and possessions in Central America, beyond some little regulation in regard to the cutting of Mahogany in British Honduras.

The treatment of the forest wealth of the United States is one long example of prodigal waste. Practically covered with virgin forests at the onset, the country has, by reckless exploitation and the cutting of every stick that was marketable, been rapidly brought to a position where her remaining resources can be fairly well gauged. No serious attempts at preserving her forest wealth have been made, present commercial gain being her first consideration ; and already she has to import from other sources continually increasing supplies to meet her native demands.

The Canadian Dominions, with a boundless expanse of forest land, also appear improvident of their wealth. No great amount of replanting or conservation is made, and already districts nearer to ports of shipment are denuded. The shipment of over £10,000,000 worth of timber annually, which is said to be the value of the Canadian export, cannot be extracted without having a pronounced effect, and already, great as are her resources, there are warning voices heard. Moreover, another cause of the exhaustion of these forests is to be taken into account, the forest fires that so frequently occur being held responsible for an enormous waste of timber, the value, perhaps largely exaggerated, being reported as equal to the amount of the annual export.

The geographical position of England so facilitates the import of timber from all parts of the globe, that no serious dearth of wood has ever been experienced, although a steady and continuous rise in the values of the imported material has occurred. Consequently, in common with other countries, England has in the first place prodigally dissipated her native forests and, later, neglected to replace the same, with the result that she is, as the statistics show, the largest importer of timber in the world, the annual value of which reaches to about £30,000,000 sterling per annum. Practically all her wants are supplied by the shipments that reach the ports, it being estimated that from the afforested land in the United Kingdom—4 per cent. of the entire area— a lower percentage than that of any other European country with the exception of Portugal—only the insignificant amount of £500,000 worth of timber is drawn.

The Kingdom, possessing only the above-mentioned and trifling amount of afforested land, has an estimated area of waste and uncultivated soil, principally mountain and heath land, which amounts to 15,000,000 acres, the greater part of which, it is stated on the best of authority, is ideal land for the growth of many varieties of timber trees. Nothing is done however. The demand for timber exists, and will continue to increase ; supplies from other countries are steadily declining ; there is land unused ; there is labour that should be employed in such pursuits, yet neither by private enterprise, by continued effort, nor by any serious attention from the State is anything done to evolve some system of the regeneration of native-grown timber. It is a weak spot in the national economy, and one which the State might well take radical measures to improve.

That timber of certain sorts can be grown has been

abundantly proved ; the native-grown Ash is incomparable for the purposes for which it is used, there is an ever-crying need for Sycamore, for Oak, as also for such coniferous woods as Larch, Spruce and other descriptions which grow freely.   None of these, however, are grown in any quantity or in a systematic way.

The question of this being done on a profitable basis is often raised, and the results in other countries abundantly prove that a direct return can be made if carried out upon an organized and business-like plan.   From statistics of State Forestry carried out in Saxony, it appears that on an area of 430,000 acres an expenditure of £330,000 was made, and from an income of £781,000, a net profit of £451,000 or 21s. per acre was realised. The rental was maintained each year, as was the capital, less timber being cut annually than was produced.

No statistics are available in reference to the reclaimed sand dunes of the Landes in the South-West of France, to which reference has been made before, but the results must be most satisfactory, the growth of the pines in these formerly waste spaces finding employment for much labour, supplying the country with a spirit which is indispensable in many industries, and practically filling the demand for pit-props in the Welsh mining areas.

INDEX

147